"BROTHERS OVER TRAGEDY"

JAIDEN LOPEZ

BOOK DESCRIPTION

Brothers of Tragedy is a heart filling, action packed adventure about the historical rage of World War Two. Brothers James and Thomas Walker are two farm hand sons that find the passion to enlist in the United States Army after an attack on their homeland. Despite their father's wishes they enlist and now find themselves in an incredible fight for their lives against the Axis forces. Join these two brothers through their journey of childhood, enlistment, training and the absolute chaos that battle has to offer, not to mention overcoming insurmountable odds. While my characters are fictional, I strived to ensure readers were met with real locations and times in which each event occurred. After eight hardworking, research filled months I am glad to finally bring this book to life. Although this was only meant for family in the beginning, I am glad that I was guided by the support of my family to publish my first book and hope all readers can enjoy. Thank you for your support.

“Brothers Over Tragedy”

Published by: Stellar Book Publishing
www. stellarbookpublishing.com

SBP | STELLAR BOOK PUBLISHING

TABLE OF CONTENTS

CHAPTER 1

THE FARM

Crop season was just around the corner; long-lasting days caused nothing but fatigue and hunger. 1929, how could we possibly know that crops were a luxury and not an American right now? Momma walks out the door with a towel, cleaning herself as dinner has just been prepared. "Come on, all days over. Why don't you come in? Get yourself something to eat?" Her name was Beth Walker, and she was a sweet lady who always had a kind heart and was as heavy in faith as my father was. Pops and I laid the plow on the crop field, stowed the horse in the barn with a fresh hedge of hay, and locked it up. My dad was a good Catholic man who never meant anybody any harm. My pa, Dick Walker, worked at the paper mill down on the road in Georgetown, Texas. He and Mama owned a farming property about nine acres long. Finding it difficult to keep up with the mortgage due to Mama not working with her tuberculosis illness and my pa making no more than seventeen cents an hour. Long days at the paper mill and crop season grew my father tiresome and distant. I would have been no more than seven years old at this time.

Education was a difficult opportunity that I could not have the privilege of having, as helping my father on the farm took most of my free time, which also caused me to be rather lonesome and socially exempt. Mama always thought that schooling was a necessity in my life, so she would hold a home class during the weekends, usually beginning around seven o'clock in the morning and ending more so around noon. She was not the smartest lady in the town, but she worked hard to give me the basic knowledge that most kids in my area were learning today. Surprisingly, I took to school as an interest, and one day, as I was finishing some homework that she had given me after her math lesson, I asked," Ma, why can't I join the kids from down the street at the public school?" She grinned and rested her hand on my shoulder as she hesitated," James, public school is for children that have time; you need to help your Father with the crops; you needn't worry no more about them, kids, you hear me?" I glared at her with disappointment as I turned from her smile and replied," Yes, Mama, I understand." My brother, Thomas Walker, was probably my closest friend, as we would often play with each other in the front yard. Often, we would start playing as friends, and then when one was too rough, it turned into a cage match. Pa drove a 1923 Chevrolet truck, and my brother and I always found ourselves behind the steering wheel pretending to drive. I would often ride passenger as Thomas took charge of driving.

Thomas was the eldest of the two of us, being about eleven years old at the time. Thomas and I said goodnight to Ma and Pa as we headed to bed, said our nightly prayers, and settled for the night. We had an early Monday morning ahead of us as we needed to begin working the field at dawn to have the most daylight possible. Thomas and I would often lie in bed and ask each other what we would do when we grew up. Faintly remembering one night, Thomas asked me, "Hey James, do you think we'll ever stop working in the field?" as he chuckled, knowing that my answer would not be to his liking. Barely capable of opening my eyes due to my lack of energy, I replied," God willing, I guess I wouldn't worry too much about it, Tom. Now close your eyes and sleep; a big day is ahead." Monday morning approached, and I could hear my father walk into our rooms with an exhausted tone, "Alright, let's go; fields aren't going to plow themselves. Come on, you, get up, and get dressed." Finding any motivation to pull myself out of bed, Tom and I got up, brushed our teeth, showered, and moseyed downstairs, where momma had prepared breakfast. Momma was always a good cook. Breakfast was usually served with over- easy eggs, toast and butter, and bacon from our pig on the farm. Pops would raise the pigs to adulthood and, when ready, drive them down to the local slaughterhouse and bring home the meat for consumption. We had a couple of trees in the backyard that Ma had grown in keeping with her small garden. Once grown, she would pick the oranges and serve us freshly squeezed OJ with breakfast. Years passed, and Thomas and I grew older, trying our best to bring our mother's education into our lives. 1930 was a devastating year for crop farmers, and we found ourselves out of work due to the Dust Bowl pandemic, crops having difficulty growing, our land's livestock dying off due to lack of proper hydration, and farmers everywhere struggling to grow anything.

Ultimately, this led to a strain on my parent's marriage and night after night, listening to the aggressive fighting that happened between them. Every morning, when breakfast came around, Thomas, my mother, and I would give disappointed looks as we noticed once the orange juice and coffee were served, our father would add to it with heavy amounts of alcohol. With my father working longer shifts at the mill and not having money for the mortgage, he went against his faith and found his savior at the end of an "Old Hemmitt" whiskey bottle. At first, he would control it, knowing that he had a family to look after, but with the Dust Bowl continuing to prove difficult, he had lost that control, and the temptation took over him. Once we were sent to bed after a low-quantity meal, Thomas and I would find ourselves no longer praying for a day off but rather for God to help us find a resolution to our parent's marriage. He and I often found ourselves lying awake at night due to the loud yelling coming from the living room outside our door, mother and father arguing over the same issue brought forth years ago. Thomas began to sob quietly and said, "James, I can't do it anymore." An uneasy feeling of emotion grew in me as I asked, "What do you mean? What can't you take anymore?" After he caught his breath, he told me," The fighting, James, the fighting, I can't be in this house anymore." worried, I explained to Thomas that only faith could get us through this. Only God knows what awaits us once the dust settles. Tired from a day's work and his sobbing, Thomas fell asleep, and the conversation was silenced.

CHAPTER 2

LOST AND HOPELESS

Saturday morning had arrived, and I knew this would be a great day to play with Tom and enjoy our rare day off. Once I awoke, I was stunned only to find Thomas was not in his bed as we usually wake around the same time, brush our teeth, and get ready together. Although this seemed strange, I made nothing of it and headed downstairs, thinking that he had just got ready early and was at the table waiting for Ma to serve breakfast. I hurried downstairs, excited for a good meal and to tell Tom my plans for the day. I was shocked to find that he was not at the table, and my mother and Father were equally confused as myself. With a worried face, my mother quickly asked me," James, where's Thomas? Get him out of bed. Breakfast is ready, and I don't want his food to grow cold." I hesitated to tell my mother he was not in the room with me. I grew the courage, told her I had not seen him, and thought he was down here with them. My father, a couple sips deep in his coffee, angrily got up, catching his fall on the corner of the table, and yelled,"

WELL, WHERE THE HELL IS HE? WHY WEREN'T YOU

WATCHING HIM?" My heart was beating faster than ever. I

looked up the stairs for reassurance and responded to my drunk father," Pa, I... I don't know; we were asleep through the night, and I didn't hear him wake." My mother, trying to calm him down, attempted to reassure us that he must most likely be outside playing in the yard and wasn't hungry for breakfast.

Seemingly calming my father as he sat down and picked up a rusty fork to eat his eggs, he turned to me and said, "Boy, get your damn brother and tell him to come eat before I whoop his ass." I hurried outside, hoping that he was out there for my sake. I began looking in the tool shed beside the house where my father kept all his field equipment and my mother's gardening tools. Thomas and I liked to play hide and seek, and this was one of his favorite places to hide. Behind the plow were torn sheets of cloth slung down a shelf with enough room for a person to hide. I yelled at Tom and told him, "Hey moron, let's go; Pa's going to beat both our asses if you don't come inside, and if I get the belt, I'm going to beat your ass," I peeled back the sheet only to find that he wasn't there. I began to get scared because he rarely hid anywhere else, and on a nine-acre crop farm, the hiding places were endless.

Mama came outside to see why it took me so long, asking, "James, what are you doing? Where's Thomas?" Thinking quickly on my feet, I lied," I found him, Ma. We are just playing hide and seek, and he said you could give his food to Pa; he's not hungry." Mama walked back inside and closed the screen door with enough visibility to see us but enough coverage to give me time to find Tom. Knowing my time to find him was short due to Pa's low patience, I began running all over the farm. I looked in the barn just in case he cleaned the stalls or prepared Betty (Our farm horse) for work come Monday. I checked each stall and behind the stacks of hay resting in the back left corner of the barn, only to come up empty-handed. Sprinting back to the farmhouse, I barged into the front door, nearly giving my mother a heart attack shouting, "Ma, I can't find him anywhere; he's not in the shed; he isn't in the barn either," Now both angry and concerned my father stood up from the table, slamming today's paper on the top and yelling, "DAMN KID, GET IN THE TRUCK, WE'LL CHECK THE TOWN!" Living in the backwoods of Georgetown, the drive was about twenty to twenty-five minutes away. Pa dropped off my mother and told her to go store to store, asking about Thomas while Pa and I drove around, seeing if we could catch him walking the street. This was no easy task because, somewhat coincidentally, the town had its monthly parade and crop festival. Due to the circumstances, the availability was low, but the local town men and women still wanted to celebrate to bring some happiness back into their lives. After about an hour of looking, we found Tom with the neighboring farm kid, watching the parade as the vehicles drove by. My father aggressively grabbed him by the arm, whispering with agitation, what the hell is wrong with you? Do you think you can just run away without consequence? Boy, you had your mother worried sick. When we get back, God himself can't save you from my belt, get in the fucking truck," With people looking at my father, he faked a quick grin, said good morning, and brought him back to the Chevy As I was walking back and noticed a group of soldiers walking behind the vehicles in their uniforms and signs reading, "Do you have what it takes?" A white man in uniform approached my father and me and handed us a brochure with Uncle Sam's face and him pointing at us, stating, "I want you." Pa quickly snatched it from my hands, saying, "The only thing that comes with that uniform is a bullet to breach it and flag for your mother," confused by my father's phrasing, I replied, "Yes, Pa," as we drove home. My father was a "Great War" veteran who had signed up at twenty. He and his high school buddies had enlisted and were stationed in France in 1917. After the war, my father returned home, although his buddies weren't so lucky. His closest friend, John Klammer, was killed in action by a bullet wound to the lower abdomen. He bled out as my father held him in his arms during a firefight with German forces. My father was shot in the left thigh and was shortly taken to a medical tent afterward. Unfortunately, that would be the last time my father saw his friend dead inside the trench, forgotten and abandoned, almost as if he was never there in the first place. For some reason, I found his hatred for the military understandable, but I couldn't resist the passion I had to be in one of those uniforms to give something back and fight for a cause bigger than myself.

CHAPTER 3

POLAND VS. HARBOR

Years have now passed, and due to continued complications with my mother's tuberculosis, her illness became too overwhelming, and she passed away in February of 1939. The passing of my mother allowed my father to increase his usage of alcohol, and within the next couple of months, my father joined her soon after. Thomas and I were left to fend for ourselves, and with my job at the mill following my father's footsteps and my brother's job, the local butchery was straining on the farm. Although we both tried to continue running the farm, we decided this was no easy task. We sold the nine-acre farm to the bank, took the earnings, and moved away from our childhood home. On the afternoon of September 1, 1939, I was listening to the radio and heard the devastating news, "After a conflict with Poland, German troops have invaded the country." I remember looking at the other workers, stunned, as we could not believe what had just come over the radio. Little did we know that this advancement in another country would cause one of the most significant conflicts in war history?

Thomas and I had moved into a home close to town, and when he got back, we both looked at each other and said absolutely nothing. As

I was walking the street to work the following week, I saw a building, soldiers walking in and out of it. Oddly, I noticed it was rather busy, so I quickly stopped in the building to see what the commotion was about. I noticed it was a recruiting station, and kids ages eighteen and up signed like they had no choice. A little over a year later, signing by choice was not much of an option anymore as President D. Roosevelt would soon pass it that men of eighteen and up would be drafted to fight America's next war. With tensions high worldwide, it seemed like this was my only option. On December 7, 1941, Japanese troops made a surprise attack on American grounds and attacked Pearl Harbor. This was enough reason for my brother and I to join the Army. We quit our jobs the following week and went to the recruiting station. Thomas was skeptical and, with worry, explained to me," James, maybe we aren't making the best decision here. I mean, what...what do we know about the military? How do we know we can do anything about it?" I waited for Tom to stop pacing the sidewalk outside the building to explain to him. I stated, "Tom, they attacked us on our land. They killed Americans in a cowardly act. Families, homes, American life itself, destroyed, we can't just let them get away with that." He stopped pacing, met my eye line, let out a quick gasp of air and said, "I don't know, Jay, maybe you're right, I'm just scared, you know, this isn't no easy path... for any of us."

I opened the door to the recruiters, and as I walked with him, I let him know we could do this together and with the help of the Lord. "Trust in me as you trust in him, yeah?" I stated. Hesitating at first, he joined me as we approached the recruiter at the office. Dressed in his tan suit and tie, he welcomed us with a greeting of the day and a question, as if he could tell us we would sign the papers today. His name was Staff Sergeant Rodriguez of the United States Army, and he asked, "Gentlemen, have you come to join the best fighting force in the world?" We both looked at each other, confused.

I shook his hand and said, "Yes, sir, we want to help; we want to make things right one person at a time." He went to the back of the office after explaining the process, and while he was gone, Thomas looked at me and said, "Ok, let's do it, let's uh, let's sign today; not like anything is keeping us here." I chuckled as I didn't think I could get him to agree with me or anything. Staff Sergeant returned and handed me the pen to sign my enlistment papers. I picked it up, and after exhaling, I put my signature on the paper and handed the pen and his papers over to Thomas. He looked at me with the face of somebody who had no choice but to trust in me. He gave me a nod and put his signature down. We handed the papers back to the recruiter; he shook our hands and said, "Congratulations, gentlemen, welcome to the Army; you ship out to basic training in two weeks; get prepared and be at the bus station by 0800 on your ship date. Best of luck to you both." Exiting the building, we glared at each other, wondering if our decision was sound. With a joking tone, Thomas looked at me and said," Well, too late to turn back now." I gave him a little laugh as assurance, and we headed back home, where we would spend the next two weeks preparing for the dangerous adventure ahead of us. on the Monday of the ship off day, my brother and I arrived in uniform, scared yet excited. Deceiving our father's wishes, we went on the bus and left.

CHAPTER 4

FORT BLISS

Thomas and I were silent on the bus ride. Two privates heading toward what seemed to be our demise, asking if we made the right decision. Luckily, we weren't far from home and were sent to basic training at Fort Bliss. Upon arrival, my brother and I thought we would stick together through basic training; knowing we were close and with the same unit, we felt safe and ready. After a rigorous entry process, what we feared the worst came into play right there, and then we were separated and transferred to different units. Thomas was ordered to serve with the 32nd Infantry Division, while I was left to serve with the 509th Airborne Division. Quickly exclaiming to Tom, I grabbed him by the arm, pulled him in for a hug, and said, "Hey, it's going to be ok, don't worry, alright? You got this; we'll see each other when possible, just stay focused and stay alive, don't pull no moron shit." He chuckled with a nervous laugh and said, "Hey, only God can control the moron shit I pull; it's in his hands; love you, brother, stay safe." We each grabbed our Army issue equipment and went our separate ways. This would be the last time I saw Thomas for the next thirteen weeks of basic training.

As I arrived at my barracks tent, I walked in nervous as I did not know anybody; everyone was a stranger, and because Tom and I were so close, I wasn't sure who I could trust and who I needed to watch out for. As I placed my Army bag on my bunk, I felt a gentle tap on my shoulder. I turned to find another private with a small demeanor and low tone. He said, "Hey, how's it going, man? I'm Bates, Ernest Bates from the Bronx." I met him with a handshake and said, "Hey Bates, I'm James Walker from Texas." He chuckled and thought of me as a horse-back riding, cowboy hat-wearing son bitch. Before I could meet every-one else, a guy we would soon know as Corporal Allen barged in to call the room to attention. He swung the door flaps back, stormed in and shouted, "GROUP ATTENTION," and coming in after was Sergeant Raymond. By the looks of him, you could tell he's been in for a minute and didn't have time for anyone's shit. Sergeant Raymond approached the middle of the room and exclaimed," Everyone shut up and put your eyes on me." Followed by silence after his command, he looked around the room and said, "When I say to do something you respond with, yes Sergeant, am I clear?" With everyone catching on quickly, we shouted at the top of our lungs, "YES, SERGEANT." He began his introduc-tion by telling us welcome to Fort Bliss, your new home for the next thirteen weeks and that he was Sergeant Raymond and he would be our Drill Sergeant. No concerns or cares about who we were back home; when we were here, we were useless, incapable of anything and here to be a part of something bigger than ourselves. "Everyone outside in two minutes for mandatory PT (Physical Training) and in the right uni-form, or I'm going to make you wish you never got on that bus." Once outside, we were again called to attention, and Sergeant Raymond and Corporal Allen ran us through our pt. session, which included a 2-mile run and vigorous upper body and core exercise. Basic training was strict and tight around the numbers. We were told when to eat, when to sleep, when to shit and when to speak. Once back in the barracks, I got to know the rest of the crew. I mainly met a couple of privates close to me throughout the training.

First, we had Private Jackson from Arlington, Alabama. Private Rusko from New Mexico. Earl McTavish from South Carolina was also there. Finally, we had Bruce Howard from Montana. Everyone had their characteristics, but we all had one thing in common: we wanted to serve and get revenge on those who attacked us. We were adjusting to our new lifestyle two weeks in, being ready for the Drill Sergeant before he walked in and having common courtesies down like we've been doing this our entire lives. Once they deemed us ready, we were finally brought to the weapons range and given our standard issue M1 Garand and ammunition to fill our mag pouches. Sergeant Raymond began our instruction class by telling us how important our weapons were in battle, stating," The rifle in your hand is the only thing that separates you from life and death; it is meant to be your friend, savior, lover." Being a farm boy, I was only familiar with my standard double shotgun, which my pa had given me for hunting. Anything with a higher caliber and faster firing rate was new to me, and I had no clue how to use it. After weapons training, I was comfortable with my weapon in terms of cleaning it, firing it, and becoming more efficient in properly using it. Surrounded by people and knowing nothing other than military bearing, I still found myself wondering how Thomas was doing each night. Thomas was older and much tougher than I was, So I figured if I could pass basic training, so could he. At the end of training, we were given leave opportunities to go home and see our family again. Inside the left building, we found ourselves to be comfortable with our NCO's (Non-Commissioned Officers) and high-ranking officers but still showed respect with saluting and common courtesies such as when a male officer walked in, greeting him with a good morning, good afternoon, or good evening sir. Our company commander barged in the door, and Sergeant Raymond called the room to attention and saluted. Colonel Theodore Morris stated, "At ease," and congratulated us on graduating Basic Training. He signed off on our papers individually and warned, "Gentlemen, basic was easy, but you must be prepared for airborne school."

CHAPTER 5

509TH AIRBORNE

Due to the high-risk danger that came with this unit, we were offered five extra dollars to our checks for hazardous pay. An increase of five dollars back in 1942 was too good to deny, so it was an easy decision to pursue this opportunity. Private Bates turned to me and said. "Hey, Walker, will you use that extra dough to buy more horses and hats? What about a new lasso to be true to your Texas culture?" With a vengeful grin, I jokingly replied, "Bates, I'm going to take your mother on a nice date to the movies, and reckon I am wasting no time and using the rest to pay for our wedding." After hygiene was conducted the next morning, we left for chow and got ready for our first jump class. Being a farm boy, the closest I got to heights was climbing the oak tree that was on our front lawn on the farm with my brother Thomas. Airborne was a new military tactic in WW2, the thought process was to transfer units by air and have them parachute into combative areas or parachute into designated areas. Looking at the soldiers near me, I could tell that everyone was nervous as the only thing going through one's mind, and mine, was what happens when both chutes fail to deploy. I thought the short answer was that you should be prepared to fight the ground because you weren't going out without swinging or trying to take the ground with you to Heaven. Just because I was at school, that didn't mean that my bearing was to be acquitted in any way. We had our first formation at 0600 hours on the following Monday.

Once hygiene was conducted, we were brought outside our barracks, and Second Lieutenant Trevor Sorensen put us at attention as we were to meet our commander, Captain Reggie Williams. He was tough but fair, always putting the unit before himself. His introduction was brief but informative as he understood the importance of our jobs as the 509th airborne division. Now, I reckoned that this would be the more difficult training of what I had seen previously. I looked at Bates and uttered, "Hey, Bates, do you think the chutes are trustworthy, or are we dead once we get on the planes?" Too nervous to laugh, he replied," Walker, considering none of us have experience with jumping, I'd like to think we are dead when we make that first jump." We began our day by heading to the jump course, where they would teach us how not to break our legs upon landing. The instructor lined up to the left of the door and had each soldier jump out of this small door frame and land a couple of feet below, with us planting two feet on the ground and rolling to our side. Uncontrolled laughing occurred as we found the process to look a little ridiculous, but the importance of the action was lifesaving. Each soldier needed five successful jumps, each increasing in height as we approached our next jump. They had us move to the runway a couple of days later. Before we knew it we checked our packs, made sure our chutes were operator ready and loaded onto the Douglas C-17 Skyline airplane capable of carrying twenty-five soldiers at a time.

Our first jump was from three thousand feet, making me uneasy because I knew it would only get worse. I would be fourth in line to jump, which was both terrifying and relieving at the same time because I would be able to get it over quicker, but I was also the first one to go. It was myself, Private Watson, the second soldier was Second Lieutenant Sorensen, and the first to jump would be our commander to make sure everyone landed safely after they jumped. We were approaching the drop zone and were given a warning.

Due to the aircraft's altitude and noise, the instructor had to yell call commands, "30 SECONDS, ON YOUR FEET, PACK CHECK, and NUMBER OFF." The first three soldiers gave their number and after me, it went down the line. Once everyone's packs were checked, the instructor ordered, "PREPARE TO JUMP... STANDBY... GO, GO, GO!!!"

Once the jump light turned green, Captain Williams was the first out. Every two seconds, the next soldier jumped, and the line shortened as each soldier gave the instructions for their zip line and dropped the terrifying seven hundred feet. The way down was breathtaking, with my heart beating faster than ever, and the only thought in my mind was, please deploy, please deploy, as if my chute and I were conversing with each other. Once I got used to the altitude, I could see the breathtaking view of the ground and the area around me. It was almost as if I wasn't in the military and was as close to Heaven as I wanted to be at the time. With ground approaching, I remembered my lander training and executed the landing technique to the best of my ability. With powerful winds, as soon as I landed, I traveled another twenty-twenty-five yards before I could gain control of my chute.

Captain immediately gave orders as I landed, "Private Walker, cut your chute and help stop soldiers from gliding across the damn ground, yes?" I responded quickly with, "Yes, Sir, moving." Once everyone had touched down, we pulled security and waited for our next instruction. The next couple of weeks would prove to be a challenge as I thought after the first jump I would have no issue with the other jumps, but with each drop, I hesitated more and more. At last, our final jump would soon come into play. The final jump would be an outstanding eight thousand feet. For this jump, I would find myself at the back of the line with Private Bates behind me and Private Durrago behind him. Once again, we were given our last warning, "30 SECONDS, ON

YOUR FEET, PACK CHECK." Accomplishing this, we would earn our wings and officially earn our spot in the 509th Airborne Division. Once soldiers began jumping, I hesitated; an intense sensation of worry came over me. Doing so could cause me to fail the airborne school and send me to another unit. Private Bates urged me to walk, saying, "WALKER, WALKER, LOOK MAN, YOU NEED TO JUMP OK, I'M RIGHT BEHIND YOU COWBOY LET" S GO, YOU CAN DO THIS, I'M

RIGHT BEHIND YOU MAN, LET'S GO." Giving me the act of courage I needed I put one foot in front of the other and made it to the front. I closed my eyes and took that leap of faith.

The jump went well at first; my chute immediately deployed, and I figured it was all safekeeping from here. However, the worst fear of any jumper came into play as my chute tore down the center, and I began to fall as gravity was taking me by the hand. Pure panic entered my body as I thought this was it. Closing in on twelve hundred feet, I couldn't calm myself to think as the plains were growing bigger and the tree shapes were becoming clearer. The wind was aggressively hitting my face, and I could not breathe. Eleven hundred feet, one thousand feet, nine hundred feet, I was closer to my demise than ever. It could have been pure coincidence, but I swear I heard Thomas speaking to me in one ear and God in the other as I remembered my emergency chute lever was to my side and ready to pull. I engaged the lever and deployed my reserve chute, and as it deployed, it was almost as if an angel's wings were spanned over me. The chute opened completely, and I began to slow down on my descent. The relief was overwhelming and tears were shed as I thought my Lord had just saved me. Due to the projection of my jump, I could not steer my chute in any direction, so I headed toward the tall oak tree planted in my direction. I collided at the top of the tree and slammed into every branch on the way down, scraping my face and tearing the open skin not covered by my uniform.

Adrenaline rushed through my body. I couldn't feel any pain until minutes later; I was just happy to have survived my final jump. Once everyone had landed, Second Lieutenant Sorensen shouted, "Alright, give me a count, make sure everyone is here." All bodies were accounted for except one. The one guy who nearly died and happened to be dangling from a tree branch. Sorensen noticed and had Private Durrago come to my rescue, saying, "Private Durrago, get that moron out of the fuckin tree please before captain notices." Durrago responded, "Yes, Sir," and headed my way. Pulling his knife from its holster, he began to chuckle and say, "Aye, Walker, I know you're a country boy, but I think the tree climbing can wait until after the mission.." Regretting life decisions and feeling nothing but anger, I rudely responded," Fuck you, Durrago, cut me down, asshole. Hey, did I ever tell you I hate heights?" forcefully putting a grin on my face and releasing a quick thank you to my Lord and savior. I rallied with my company and apologized to the Lieutenant, "Sorry sir, I... had an accident in the sky." We got accountability and proceeded to finish the mission. Later that evening, we gathered at the Officer's only club, which was exclusively a building for High-ranking officers.

However, this was a special occasion as we had just graduated from airborne school. We wore our Army AGSU's (Army Green Service Uniform) and had drinks and food to celebrate. Laughing and sharing stories ended quickly as Captain Williams called the room to attention for the company commander, Colonel Morris, He allowed the room to be at ease and gave his congratulatory speech "Gentlemen, you should be proud of yourselves as you had just accomplished one of the most dangerous and terrifying schools the Army has to offer, you put your lives at stake for a higher cause. Understand that your contribution to go airborne and land on the enemy head-on is much appreciated by the Army, and you have all earned the right to be a part of our airborne division. Gentlemen make one single filed line and... come get your wings." The Colonel handed out the wings and shook everyone's hands. Once my opportunity came, Morris gave me a quick chuckle and said,

"Private Walker, I heard about your little accident on the final jump. I'm glad to see you made it, son. Congratulations," and I shook his hand with my wings in his. With a grin and a sense of embarrassment, I replied, "Thank you, Sir, glad to be here too." I wrote a letter to Tom telling him of the good news and a couple of days later got his response. "Hey James, congrats on your wings, brother; all good on this end; training's been a sumbitch, but I'm tough and was able to pass. I was recently promoted to Corporal, and you won't believe it; it gave me more responsibility. Should have stayed home and stuck with the butchery but I couldn't let you have all the moron nonsense yourself, love you brother, stay safe" happy to hear from my brother, I put the letter with the ammo tin I had placed under my bunk for safekeeping, took Private Bates and Durrago for a smoke in the smoke pit outside, and discussed issues we knew of. With President Franklin D. Roosevelt declaring December 7, 1941." The day we live in infamy," we knew that tensions were only increasing as we had declared war on Japan.

Soon after, Germany and Italy declared war on us. We knew that our unit would soon be called to action, and our first live jump was only creeping our way faster and faster. The United States had officially brought forth twenty-five other countries to create what would then be known as the United Nations. We knew that we weren't in this fight alone, but we had a lot of countries relying on us for support, and we had to deliver. Private Durrago puffed his cigarette and jokingly stated, "Ya know if I had the chance at Hitler, I'd put my gun up and use my knife to scalp his ass slowly and painfully." Bates and I both stopped mid- puff to laugh at his insane explanation of killing Hitler. Meeting each other's eyes, we couldn't believe what this crazy man had just said. Bates tapped off the dry end of his cigarette and put the joyful mood to a halt, and he changed his tone to a more depressive state and said," Ya know I hear about' fifty thousand Jews had bit the bullet, done damn near killed them all. Put the lucky saps in a prison camp, torturing them daily." Durrago inhaled another puff and said, "Shit to be alive and tortured. I don't know if I would say that's lucky, if you all ask me, sounds like the dead Jews were the lucky ones." I leaned over to put out my cigarette and stepped over to sit on the bench that connected to the table. I shook my head as I turned to look at the night sky and uttered," Well when the time comes to an end that fucker, I reckon us boys will get our chance and they'll be sorry, I know that for sure.." Giving a more accurate number, it was determined by "The New York

Times." that number would unfortunately be a lot higher than we anticipated. The projection was determined that around one million Polish Jews had sadly lost their innocent lives due to their religion by February 19, 1942.

Utter silence grew over us as we returned to our cigarettes in awe of what was just discussed. Second Lieutenant Sorensen left the Captain's office and headed our way. I called the group to attention and rendered a hand salute. Sorensen asked to bump a cigarette off of me and asked Durrago for a light. As he puffed, he stated, "Thanks, boys; I take it you heard about those poor bastards in Poland, huh?" We each gave him a silent nod and said," Yes, sir." Private Durrago asked the Lieutenant," Hey, sir, when do you think we'll have a shot at that son of a bitch? I mean respectfully, sir; we can't let him get away with this." Lieutenant Sorensen inhaled another puff, took a seat at the table next to me and hesitated with his answer, almost as if he knew that day would never come. Not answering the question, he finished off his cig and stood up from the table. Walking away, he stopped, turned to us and said, "Focus on your training for now, boys. Wars are not going anywhere, and I believe that day isn't too far out." We again gave him a nod as we knew that was a tough question. We pressed our cigarettes against the table and turned in for the night. Wanting to stay up later as we had our weekend pass the next couple of days, I was afraid my eyes wouldn't let me as the long and stressful days were too exhausting for me, and I decided to head to get some shut-eye. As I woke up the next morning, I finished another letter to my brother Thomas, nothing too exciting, just writing what we had discussed last night and telling him I should keep strong. I spent most of my day with some self-PT and had some running time, so I took advantage of that and watched a feature film later that night.

CHAPTER 6

GERMANY INVADES STALINGRAD

Our company had just finished morning chow on August 25, 1942. Our company commander had just barged through the door flaps and said he had urgent news and needed all eyes on him. He looked around the room and said, "Gentlemen, we just got word that Germany has advanced into the U.S.S.R and has declared war. The invasion happened a couple of days ago, and they have begun to fight back against them fucking krauts." He said urgent readiness training needed to begin immediately, and we needed to prepare ourselves for the worst. Hitler intended to destroy what he thought was Joseph Stalin's (Premier of the Soviet Union) Jewish regimen. Hitler warned the Soviet Union that they needed to stop preparing for war against Germany, or he would invade to cease the problem.

From the beginning of this conflict, Russian forces could not slow down German Wehrmacht's forces. Stalin's forces were projected to have around two hundred thousand losses but successfully held off the German force.

Everyone in my regiment was rather worried about the Russian forces being able to hold back the German forces. I remember Private

Durrago catching me before I went to sleep that night and telling me, "Hey, Walker, you think the war will end of them Soviet fuckers kill them all?" Slowing waking back up, I groaned as I brought myself up from my bunk and leaned against the wall to engage in conversation, "To be honest, Durrago, I reckon we'll be in the fight no matter who kills who; the Soviets win, and we're their next target, shit say they did kill each other, we still have Italy and Japs to worry about. They haven't given us all this training for shits and gigs. Catch some sleep. Tomorrow is a new day, and we got two jumps, so focus on pulling your chute and not dying." He responded sarcastically, saying, "You mean like you, did you forget your reserve, or are you just an idiot." Not responding to him, I gave him the middle finger, laid back down, and went to sleep. A couple of days later, we heard that Germany had launched their Luftwaffe (German Air Force) attack and had sunk a couple of Soviet vessels in the process. Unfortunately, by September, the Luftwaffe had taken over the skies of Stalingrad. Stalin had become desperate for troops and eventually asked Russian civilians to take up arms and join the fight.

Now, this wasn't as wild as Adolf Hitler forcing men and women to join toward the end of the war, but Private Bates and I wouldn't be surprised if he started a draft just as the United States did earlier in the war. Unable to keep the German forces out of the city limits, fighting raged into the city, and the Soviet Union began to use heavy sniper fire on German forces from rooftops or hidden balconies. Russian generals Georgy Zhukov and Aleksandr Vasilevsky organized Russian troops in the mountains north and west of the city. From there, they launched a counterattack, famously known as Operation Uranus. They constructed a battle formation around the city, almost like a defense circle around Stalingrad. All seemed lost until the winter front came into play and allowed the Russians to take advantage of the Germans. Germany was not used to the cold that overwhelmed Russia like the Russian forces were. Due to their low-calorie intake and not being able to become accustomed to the freezing temperatures, German forces became exposed to hypothermia, and it wouldn't be long before troops began to fade away due to their conditions.

During the invasion of the Soviet Union, also known as Operation Barbarossa, it would be gathered that about one hundred thousand Germans had succumbed to frostbite, with nearly ten percent needing amputations. The German forces' food supply and ammunition began to deplete in rapid amounts. Even with the inevitable loss of the battle, Adolf Hitler still kept troops in the fight and slowly watched them die off to the Russian defense. Hitler did not believe in defeat, much like the Japanese culture, and the only way to end a war was to die in it for the countries good, even if the Germans didn't completely agree. I often thought, "Why would a chancellor take more pride in killing off his troops rather than save them for future needs? Was it to save the embarrassment of loss? Is it true that Hitler doesn't make any sense? I reckon I'll never truly understand the bastard." Russian forces had retaken Stalingrad's lands and successfully defended its territory against the German invaders by February 1943. Not to mention, it had captured nearly one hundred thousand Germans in the process.

Hitler finally acknowledged this as his first loss, this helped the allied forces of Russia, France, Britain, and the United States, and gave them the confidence they needed to fight against the Axis powers. Months followed, and as I was resting on the first day of my weekend pass, Private Ortiz ran in the door flaps of the barracks in a hurry, almost as if he was going to tell me the war was over and we could go home, instead he raced to say, "Walker, walker, boys, boys, aye let's go Cap wants us to meet outside 5 minutes, let's go, get dressed and form up outside." He steps out, noticing that he couldn't hear any movement from inside the barracks. He walked in once more with frustration, "Aye, get y'all asses up; captain wants us to form up, get the fuck up!" Hesitant to get up from my comfortable sleep, the remaining men in the barracks got up, put on our pt's. and formed up with the rest of the platoon outside. The Captain headed to the front of the formation and stood before the platoon, "Lieutenant, are all men in formation? Have you gotten accountability?" Lieutenant Sorensen rendered a salute and, once dropped, said, "Sir, all men accounted for, Sir Do you know why we were called?" Looking at the men behind him, Captain Williams moved closer to Lieutenant Sorensen and whispered, "I think we just got word about the Soviets, but I wouldn't get too excited; I think we are about to do our jobs for once, god willing, I suppose, let's wait for company commander alright?" Sorensen responded, "Yes, sir," and went back to the position of attention. We must have stood there for damn near Fifteen minutes before finally seeing Colonel Morris approach our formation. He had a smile on his face, almost unsettling and what seemed to be a sign of good news, but then again, this is the Army, so how good can any news be? Morris stated, "Good morning, gentlemen." We all gave the Colonel a greeting of the day and awaited the anticipated news, "Men, we had just received word that the German forces failed to take over the Soviet Union and Russians had successfully defended Stalingrad; hail to mother Russia, right? Now while this is good news to us because they nearly killed all those fucking krauts, Germany has only enhanced their defense and has become more aggressive with their attacks." Private Ortiz was next to me in the formation and leaned over just enough not to be noticed moving while at attention and whispered, "Hey Walker, isn't no way they killed all them krauts, right? I mean, half the country must have fought for them."

Trying not to engage so as not to get yelled at by our CO, I softly spoke to Ortiz, "Man, I hear it was the cold like they couldn't handle it or something; heard Hitler damn near starved them just to keep them in the fight, now shut up before Morris gets on our ass." Colonel Morris concluded by telling us the news and gave orders to Captain Williams, "Captain, that's all I got for now. Get your men to chow and have they ready for free fall come Monday morning, alright?" Captain Williams rendered a salute and replied, "Yes, sir." As Lieutenant Sorensen was dismissing the formation, I heard a voice from the crowd asking for Walker. Unable to see past the other men, I yelled, "I'm Walker," only to realize Williams was calling me to the front. I rushed over, and he told me Morris wanted me to report to his office right after chow in full uniform. Too nervous to eat now, I sped through chow and headed straight for the Colonel's office. I knocked on the door leading to his office and opened the door, "Sir, Private Walker reporting to you, sir; Captain Williams said you wanted to see me?" He told me to come in and take a seat.

As I was sitting in the chair, he began to pour a glass of whisky, asking if I wanted some. Due to my father's issues in the past with alcohol, I respectfully declined. Morris said, "Didn't take you for the non-drinking type, let me guess, abusive father? He laughed as he stopped pouring my glass, "No worries, son, I respect a man that can hold against the temptation of delicious sin," He began by telling me that he was impressed with my drive and facing my fears when it came to jumping and that he had noticed I had put some good effort into sharing the courtesies of my platoon. He told me that the welfare of his people was his duty, and he wanted to let me know personally, "Walker, you have a brother with the 32nd, right? Tim, Tony, something with a T, right?" Not wanting to correct an officer, I felt obligated to do so, seeing that it was my brother, "Thomas sir, Thomas Walker, and yes, he's with the 32nd, is...is he ok?" He took a sip of his whisky and responded, "He's breathing, son, but I'm afraid during his tour to Africa, he received a bullet wound and was administered to the nearest triage hospital to look after his wounds. Don't worry, he's alive, but I thought you'd like to know." Thinking only about my brother's well-being, I was at a loss for words for a few seconds. He sat on the front of his desk, resting his arm on his lap, and said, "Now, seeing that you share a rather excitement for war, I, as well as the president, have seen fit to promote you; I'm bumping you up to Sergeant ok? Stand up, son." I stood up quickly in a hurry to accept my new responsibilities, "Yes sir, thank you, sir," he shook my hand and told me that the platoon had to look up to me now, and I needed to be a prime example as well as help out the Lieutenant and Captain.

He dismissed me, and I saluted and walked out of his office. I immediately went back to my barracks and began writing letters to Tom. I asked about his condition and wrote that the moron shit was supposed to be under control while we were separated; I guess that wasn't possible without my supervision. I asked him if he had also received word regarding Russia and if Africa was the next defense for us, not knowing that he was over there in the first place. I told him about my promotion and new tasks that lay before me and how I was technically his boss, knowing damn well he wouldn't listen to anything I said. We all talked in the barracks for a while, and the guys congratulated me on my new promotion. We continued to talk about Germany and its failure to overtake Russia. Although it seemed like the war was ending, it was far from over.

CHAPTER 7

NORTH AFRICA

Although it seemed as though the invasion of North Africa was nearly over, it didn't save us from getting our first chance at combat. We were called to yet another formation to give us the heads up that we were being called up to ship out to Africa. We were to leave in the next couple of weeks, and we were going to parachute to the designated area and land in our objective. Privates Ortiz, Bates, and Durrago were extremely nervous as this would be their first combative mission, as well as mine. Still, as a leading NCO, I knew I had to reassure them that we would be fine if we stuck to our training and remained focused. Of course, by knowing this, I was not confident in the matter, but I could not let that show; otherwise, we may as well go AWOL now.

509th Airborne troops landed in Africa on April 23, three days before the attack against the Axis forces. With allies on our side, this was our best chance of pushing back the German and Italian forces. We began packing on April 25, 1943 and readied our gear. We were briefed on our mission. We would jump from our C-17 airplane , and land on the fronts of the Axis forces to act as a divider and defend from any enemies. Having almost zero knowledge of what to expect from these forces and only understanding our capabilities as to how we were trained, we did not take this lightly and underestimate our enemies. April 26, 1943, today would be the day that we were a go for the mission, and by early morning, we were loading into the planes and getting ready to fall to what seemed to be our death. When given the green light, all the C-17s began to take off from the runways; I would be responsible for seeing that every soldier successfully exited the plane without issue and joining them in the back.

Storming toward our objective, we immediately began taking heavy fire from enemy artillery and gunfire from the ground. With nearly forty airplanes in the air, it wouldn't be long before artillery projectiles would hit the plane twenty yards to the left of us and cause it to become engulfed in flames and start to barrel toward the ground. Instantly taking fire on our plane, I could do nothing but pray to God that he let us make it to the drop zone in time. Captain Williams began shouting commands immediately, "EVERYONE GET DOWN, STAY LOW AND BE READY TO JUMP IN THIRTY SECONDS." We began to check each person's pack and ensure we were set. I began to shout, "EVERYONE UP, GET READY," the green light at the beginning of the plane turned on to signal that it was time to jump. I opened the door and began sending people out of the plane, "GO, GO, GO!" Men started to move down the line, and while taking fire, heavy caliber bullets began to pierce through the plane's wall and struck a private as he was moving down the line, causing him to fall over and trip the next soldier in front. I ran to the back of the plane and began to move his corpse out of the way to clear the aisle so the line could continue. Once everyone was out of the plane, I wasn't far behind, and with who remained, I successfully jumped.

Unfortunately, seven other planes joined the casualties, reducing our numbers for a defensive force once we hit ground level. We fell quickly, purposefully, and hurried to the LZ; once we landed, we got accountability and cut our chutes loose. German and Italian forces were about eight miles away, so we regrouped and pushed that way to meet up with the 32nd Infantry, which had a defense line already anticipating the attack. Our journey to them was quiet, and when we met up with the 32nd, Captain Williams met up with the acting commander, Captain Ernest Wright.

We approached the unit, and Captain Williams asked, "Hey, where's your CO?" The private pointed him to the nearest tent and told us to rally up with the others behind barricaded walls made of sandbags. He walked to the tent where Captain Wright was waiting to brief. They shook hands, and Wright began to explain, "OK, Cap, we received word from a runner that Krauts and Italian forces were heading this way and should be here by morning, we have this shitty defense line at the ready, and we are waiting for them to come." Captain Williams began to ask, "What are we looking at for troops and ammo?" Wright responded, "Honestly, we're pretty shit for ammo, and they took out damn near a battalion a couple of days ago; word must have reached out if they are sending more of them fuckers." After getting the brief, Williams ordered us to set up camp and build some trench holes behind the tree line surrounding the line. We dug in for the night and began to eat the shit meals in a can that were issued to us. "Raw meat and crackers to feed us; we're sure to win now," said Bates after we dug our hole. We heard Second Lieutenant Sorensen tell us, "OK, boys, get some rest, pull security and switch every two hours, Roger?" We all responded, "Yes, sir," I told Bates I would take the first watch.

Throughout the night, we could only hear gunshots and artillery cannons going off in the distance. Not exactly the sound of silence, and sleeping was nearly impossible. I woke up due to the noise and conversed with Bates and Durrago. I asked if anybody had received any letters from home, and Durrago responded, "I got a letter from the misses, saying baby number two is on the way; we play this right and might be able to see the kid before Christmas; what about you, Sarge?" I grabbed a couple of unopened letters from my pouch and began to open them to see what Thomas had written to me, "Just some letters from my brother; sounds like he's here with us, but I'm not too sure where, isn't like he'd tell me anyhow." We turned in, and Private Durrago had the final watch. I couldn't help but think what I would tell his wife if he didn't return. It turns out my thoughts came to reality at dawn as we heard something climbing out of the pit. It was Private Durrago, and he went to take a piss. He woke me up and said, "Hey Sarge, going to take a leak, be right back," I said OK and told him to keep his head down. An utter silence grew over us until the sound of a sniper roared over us. Bates and I had heard something smash back into the pit, and when it woke us up, we noticed that it was Durrago falling back into the hole. I immediately began shouting after realizing what had happened, "DURRAGO, DURRAGO!!!" Lieutenant Sorensen told everyone to get up as we had been receiving contact from the other side behind the tree line in a small trench that Germans had been building the night before. Due to the gunshots and artillery guns, we did not hear them as they kept quiet and moved slowly. Durrago had taken a sniper round through the neck and began bleeding out profusely. I commanded Bates to start firing back, and he loaded his M1 Garand and shot back at the German forces. Quickly acting, I started to put pressure on Durrago's neck, almost to the point where I was nearly inches inside the entry wound, trying to keep the flow of blood from exiting the wound. "MEDIC, MEDIC, I NEED A FUCKING MEDIC

OVER HERE NOW, MEDIC." Duraggo, gasping for air, was seconds away from dying, and a medic crawled to our pit and began administering aid. "Stay with me, Durrago, come on, stay with me; you're going to be OK, fuck, stay with me, man." The medic began to push me off of Durrago so he could give him aid, but because he was a close friend, I could find myself to stray away, angrily the medic said, " Sarge, you must move away, he's not gonna make it. Sarge, Sarge, you need to step away, now!; he won't make it. Sarge, Sarge, you need to step away." Bates, ducking low after emptying his mag, began to shout in my direction, "Sarge, Walker, we need you up here, man; he's dead; we'll come back for him. Let's go!" Durrago took his last breath and died right there in my arms; nothing but an echoing sound of gunfire entered my ears as I paused for a second, collected his dog tags and said, "I'm sorry, Durrago, I'm so sorry." Collecting myself, I grabbed my weapon and met Bates and the other firing soldiers behind the barricaded wall of sandbags.

Captain Williams noticed the German forces had a trench-like system on the other side. He crawled behind the barrier to meet up with Lieutenant Sorensen and I. He said, "OK, here's what we will do, Walker. You're going to take five men and push left and flank these fuckers, LT you go with them, I'll take some men to the right, and we'll meet in the middle, alright? On my signal, pop smoke near the tree line, wait for a cease-fire, and we'll put down some cover fire on the fuckers. Move quickly and stay low; announce friendlies when we meet, Roger?" Captain Williams ordered, "OK, on three pop smoke, one, two, three, now, go, go, go!" Bates and I popped smoke and began pushing left into our side of their trench. We saw Germans rushing over their wall and began to fire. I hit two in the chest, and they dropped, and Bates fired at another, causing him to fall as he was going over, hanging his body over the wall. We paused right before an entryway into their trench, and Sorensen told us to grab a grenade, "OK on my count, here we go, pull pen, wait, and wait, NOW!" We threw the grenades and ducked behind the wall. As soon as we heard screaming, we pushed into the main entrance. LT Sorensen dropped two that were concussed and shot at the others, pushing further back into the trench. Heavily breathing, all I saw once we entered were three German soldiers with missing legs from the blast and another soldier resting against the wall with an entry wound through the top right of his head. The immediate sense of realness hit me like a truck, and I couldn't believe we were this close to the enemy.

We announced friendlies as we heard troops coming from the other direction. It was Captain Williams. We gathered a quick count, and he had lost two men on his way over, both subjected to bullet wounds. Lt updated the Captain, "Sir, clear left, but they ran back through this trench; I can't see past this wall; we got two sides, one left entry and another on the right; what's the order, sir?" Other soldiers grouped on both sides of us, and we were ready to move again. Williams ordered, "OK, the same thing, you hit left, and we'll hit right; everyone reloads and follows Lt., Lt on you, ready, go!" We began slowly approaching the left side, where an awaiting German MG42 shooting in our direction pinned us down. One soldier met this weapon by surprise as he turned the corner and took nearly seven rounds to the chest, pushing his body back to us. We pulled him back and handed him over to our medic, who put him on the litter and pushed him back to our side of the barriers. Lt asked, "Anyone got a grenade?" A fellow private pulled the pin and tossed it to me; I threw it over the ledge into the open German bunker. The grenade did not make it across, so it landed on the higher-leveled ground, and the blast went right over. They immediately began shooting another load of rounds our way. Private Jenson was our radio guy, and Lt called him to his area. "Get the Captain and tell him we are pinned down by heavy fire and can't push left." He radioed, "Top one, Top one, this Eagle two; I say again this is Eagle two; we are pinned down by the heavy fire, lethal explosives ineffective, I say again, lethal explosives ineffective, we require assistance, how copy?" We reached our weapons over the ledge and began free-firing in their direction, seeing if we could get lucky and hit the MG. Captain Williams returned on the radio, "Heard you, Eagle two; we are advancing right and will flank their position. Keep them distracted so we can flank." I ordered Bates and the other soldiers, taking cover, "Top's flanking; give them cover until they can reach our position."

The next thing we heard was a grenade blast, and we began to hear screaming in the bunker; it was not a lethal kill, but the Germans were dazed; we heard Captain yell at us to push forward, and due to the close distance, we could not pull our weapons up to fire so we began executing in hand-to-hand combat with the Germans. Bates took the soldier to the right and began to strike him with his combat knife, dealing two strikes to the chest and one to the throat; I hit the soldier on the MG with the buttstock of my weapon, and as he fell, I started to strike him until the threat was neutralized, as I looked up I could see the rest of the pushing soldiers striking and killing off the remaining Germans in the bunker. We continued to clear the remaining trench and found three Germans waiting on their knees with a white flag in hand. Commands were given to remain on the ground and put their hands behind their back. Bates walked over to one of the soldiers and delivered a strike to his face saying, "Fuck you, Kraut, get on the ground," and tied his hands, and we escorted them back to our side of the tree line as our POWs (Prisoners of War). Once the trenches were cleared, Captain Williams said, "OK, Lt, get a count of wounded and KIA, grab any classified information and replenish ammo." We picked up our casualties and rushed them back to our side. We grabbed all German weaponry and destroyed anything they could use against us. Come nightfall, as we pulled security and rested in our holes, I made a quick stop by Durrago's dead body, which was covered by a medical blanket. I peeled back the sheet to see his wounded body and fell to my knees. I couldn't believe that somebody that I had trained with, hurt with, and led into battle was gone in a matter of seconds, not even capable of firing a single round, with almost no warning and so much waiting for him back home. I grabbed anything on his person that was considered personal. Behind his magazine pouch, I saw a bloody paper tucked behind his uniform pocket. I pulled it out and opened the blood- stained note and saw that it said, "To Ann and Susie, from me." I sat back on my bottom and opened the letter. It said, "Hey, beautiful ladies, it's me. I just wanted to let you know that we have a mission in a couple of days, and I'm unsure when I'll be able to write again. Don't worry about me; I'll be home as soon as I can; After all Susie, I can't let Mama be the one who teaches you to ride that new horse, ha. Ann, I can't wait to meet our new baby boy. He must be getting so big now; I want to squeeze his little bones badly. I'm sorry about the poor timing, but I want to be there for everything when I return home. Pray for me, ladies, you all are what I'm fighting for, and you keep me going. See you all soon; I love you all very much." Bates came over and sat next to me, asking, "What's the letter say?" Putting the note in my pocket after trying to clear some of the blood off, I told him it wasn't anything that could help him now.

Bates pulled out a flask that he had kept in the loose spot of his boot and gave me a swig. "Drink up, Sarge, to Durrago." I grabbed the flask from his hand and whispered, "Yeah, to Durrago," and took a couple of gulps to flush out the dirt and blood caught in my throat. Lt. Sorensen crawled in our direction as he checked up on all the soldiers; he rested his back against the sandbags and sighed as he sat near us. "Helluva day, huh boys? One helluva fuckin day." He laid his hand on my shoulder and gave his condolences to me, "I'm sorry about Durrago Walker; he was uh... he was a good man, had a lot of potential." He dropped his hand off my shoulder and laid his head back to release another sigh. "You know we have another set of Krauts coming, should be here by morning; we're going to use the son of a bitch's trench to our advantage and push up; word is they got artillery aimed at the skies to take our boys down in the air. The mission is to take out the guns and kill or capture any German defense, so rest up, switch out every two hours and be ready for tomorrow, night boys." He stood up and kept low as he headed back to his hole. We responded," night sir." I caught his attention once more before he was out of whispering distance and asked, "Sir?" He halted and turned my way once more," What's up, Walker?" I turned from looking at the ground and said," Lot more sumbitches going to die tomorrow, huh?" He hesitated, answered the question, and said," Get some rest, big day tomorrow." We turned in, and the next morning arrived.

CHAPTER 8

GERMAN ARTILLERY

The next morning, we were once again awakened by the sound of the German Artillery guns cracking off rounds in the distance. We were all ordered to ready our gear, load weapons and be ready to move in five minutes. Troops gathered in the German trenches and we pushed forward into our next firefight with the Axis forces awaiting our arrival. We refilled our ammunition pouches and proceeded through the trench. With every step we got closer to the guns and could hear them firing off rounds about every seven seconds. From the sounds of it, there were about three separate guns and you could just barely see the ends of the barrels sticking out above the trench which gave us the assumption they were spread about fifteen feet from each other. Captain Williams halted us in place and gave orders to Lt Sorensen to pass down the line, "ok we're going to split again, Lt you take the left and we'll take the right, make sure your demolition man is near the back, we can't risk losing him and dropping our TNT, so on my count we are going to move quick and with fuckin purpose Hooah?" Acknowledging the orders I went down our side of the line telling the soldier behind me to check gear and weapons and pass it down the line. The last man was checked and we were ready to advance. We couldn't risk going over the trench and running on level ground as the weapons were behind the tree line and we couldn't see enemies hiding behind it so if we were to run who's to say we wouldn't lose half the battalion trying to rush so we moved low and quick through the trench.

I gave a thumbs up to the Lt and he signaled to the Captain and before we knew it he gave the green light and we advanced. The firefight began almost immediately once more as MP40's and Kar98 sniper fire was headed our way. Under fire we lowered behind the barricade and every couple of seconds poked our heads up enough to fire rounds down range. Bates killed nearly seven to eight Germans on the other side within seconds, his M1 clip emptied and he reloaded the next set of rounds. A young Private by the name Jeremy Taylor peeked at the wrong time as a round pierced through his helmet knocking it off just for another round to enter through his skull and dropped him next to me. Carrying on we were able to get into the last part of the trench before the open area of which the guns rested. Noticing that we were on the other side the German force turned their attention our way with two more MG42's firing in our direction. Panicking and thinking they wouldn't make it past, two soldiers rose up and began running to the nearest cover, they were caught in the crossfire and took rounds to the side and were thrown to the ground and were KIA in seconds. We radioed the Cap, "Top one, Top one, this is Eagle two, what's your status, over?" The radio man on the other side responded back, "Eagle two, this is Top one, we are caught in heavy fire, as of now three KIA's and Captain took a hit to the arm, he's alive but down for the moment while medic's render aid, keep pushing we are just outside the gun walls." Lt gave a good copy and we made ready three grenades to toss over and potentially hit the soldiers on the other side. One grenade made contact on the right side giving our troops the advantage to push in and find cover on the other side. I told Bates, "Ok they made entry on the right, provide cover fire and we'll push in, roger?" Bates acknowledged and as he stood up to fire he was met with an MG round to the shoulder pushing him back behind the barricade, "AH, I"M HIT, I'M HIT." Immediately lowering him to the ground I shouted for a medic, "You're ok Bates, just an exit wound ok, went through your shoulder, you'll be alright, here put pressure." I gave orders to the medic to give him morphine and pushed him to the back of the line. Lt shouted as to hear over the gunfire, "OK EVERYONE COVER FIRE AND PUSH FORWARD!" We all rose at once unleashing hell and emptying our magazines and managed to kill off damn near the whole platoon of Germans firing at us.

Once cleared we advanced toward the guns and our demolition man planted the TNT on each of the three flaks. "GET BACK, GET BACK, Take cover," he shouted as he was ready to blow those guns to hell. We all ducked behind the barricade and the charges went off destroying each weapon. Mission accomplished with minimal casualties and damn near losing Bates we got accountability of troops, weapons, ammo, and KIA's both friendly and enemy. Lt Sorensen went to check on Captain Williams, "Sir, are you good, do you need medical?" The Captain tried to save face and said, "I'm good Lt, just clipped a wing but I'm fine, what's the status?" Lt gave him an update on the destroyed weapons and casualties on the field. We radioed to command to give the news and set up camp for the night as we were to meet up with what I would soon know to be the another platoon awaiting at what would be the final advancement to take out the German and Italian forces. We dug another set of pits and I rested with Privates Ortiz and Bates and a few other guys that had fought that day. Sitting down I told the boys, "Alright y'all good day, get some chow in you and sleep while you can, Bates make sure that cast is tight, might need everyone for what's coming alright?" Bates was a tough sumbitch and told me that it's going to take more than a round to keep him out of the fight. "I'll be able to fight Sarge, don't worry about me." Word got to the company that the Axis forces were preparing for their final attack in North Africa and we knew that with our resources we were at a slight disadvantage.

We received information of a company nearby that was defending the planned destination of attack and with the artillery weapons cleared and destroyed we made our way toward friends. Once we arrived, Captain's Williams and Wright made their way to the Commander's tent to get in contact with the acting commander First Lieutenant Olsen. In front of the tent flaps there was a Corporal standing guard and Williams said, "Corporal, where's the commander?" Dropping his salute he said, "Inside sir, but it's not the captain, its Lieutenant Olsen sir." Looking at each other with Confusion they both walked inside the tent and Captain Wright yelled," Aye, where's the acting commander at?" Due to him being the highest rank in the room they all stood at attention and the first private in the room pointed toward the back where a large geographical map of Africa was placed on a table. They made their way toward the back where they would find a young Lt who could not have been more than twenty-seven years old. Captain Williams immediately began asking, "Lt, where's your Cap? Why is he not here briefing me of the shit storm?" The Lt looking as if he was on the front lines for the last three years responded nervously, "Uh sir, he uh... he didn't make it in the last advance by the Germans, sun of a bitch took hot brass to the face, caught him dead right then and there, like that we were short a leader so I took over."

Sighing and rubbing both hands down his face you could tell that this was the last thing he needed to hear at the moment. He laid his weapon down, slanting it on the table. They both took off their helmets and Captain Wright pulled out a small bottle of whiskey and told the Lt, "Here Lt, pour a couple glasses, take a seat, sit, relax, and tell us what's coming." Pouring the whiskey with shook hands, nearly spilling it everywhere he began explaining what we were up against on this next attack. "Right, um, so, yeah we uh," not capable of spitting words from his mouth. The Cap told him to slow down, breathe and asked what we were looking at in terms of forces, "How many men, any vehicles or tanks? What about weaponry?" The Lt finally caught his breath and said, "Yes sir, we expect around three hundred men, both Kraut and Italians. We sent a scout out yesterday and he came back saying he could only see transportation trucks but had MG's and a lot of sniper fire so y'all need to keep your heads down." Setting his glass on a table behind him the Lt opened up the map further and pointed toward the ground of which we expected to fight, "We should see Kraut here, here, and here, if we can take out them bastards here we may have a chance, we hear it's mortar units and if they fire like hell on our position I can't see any of us making home." Captain Williams looked at Wright and explained to him that we should separate our troops evenly between the three of us and each leader takes his guys to neutralize each mortar threat. He looked at the Lt and said, "Lieutenant you're going to take Sergeant Walker and his men and you guys will push right toward weapon Alpha, Wright, you take your men and push toward Bravo and I'll take Charlie and we will rendezvous and point David to get accountability of all soldiers and weapons and we will hold off them fuckers trying to take Tunisia, understood?" Both meeting each other's eyes they nodded to show they understood each other's assignment. Williams added, "Alright when you get to the points, lay on some heavy fragmentation and blow those fuckers to hell, copy," They nodded once more. They finished the briefing by telling each other that we should hit them hard and heavy in the morning. April 26, 1943, the morning of the final invasion of Tunisia, this was sure to be our biggest battle yet. We tucked in for the night one last time before all hell broke loose tomorrow.

Again, it was myself, Bates, and some new privates that jumped in the previous night before, they rallied up with our company and were assigned to my leadership. They couldn't have been more than twenty years old. One caught me as I was drifting off to say, "Hey Sarge, you think we'll make it tomorrow?" Keeping my eyes closed to rest but still engaging in conversation I exhaled and murmured," Private I'm not

God alright? I can't tell you who's going to make it and who will see their last day tomorrow. Just do as you're told, be smart and calm and we'll see if you're blessed or not. Sumbitches aren't easy but you listen to me and your leaders and y'all be alright." He put his head up against the dirt that was acting as a wall and said, "Roger Sarge." The morning hit and we packed our gear and made our way to the defense point. We got the word from a runner once he returned to our side that the Axis force was approaching our position and that we needed to make weapons ready and double check all gear and ammo. This would be our biggest challenge yet as the troop numbers nearly quadrupled what we had fought previously. Privates near me were looking up and around and then Bates said to me, "Aye Sarge, do you hear that?" Turning my head side to side and up and down we could all hear the sounds of enemy troop footsteps and marching as well as a German rhythm in their voices. Captain Williams shouted, "Alright men, this is it, we hold them and we may take Africa. Load up and standby, nobody fires until I say so, roger?" Everyone acknowledged and took their weapons off safe and made ready for the fight only a few hundred yards out. We had eight Browning M1919A4 machine guns posted on the grounds behind a barricaded wall with each weapon consisting of nearly three hundred rounds. One shooter and one feeder on each weapon ready so if the shooter goes down one is at the ready. Twenty feet away from them were four soldiers holding M1A1 Bazookas, although powerful ammunition for these was low so we needed to use them at the right time.

Germans and Italians began crossing the tree line making them visible and targets for our Browning. Captain Wright gave the order and the first round of an M1 Garand blasted through the soundwaves like a firecracker in a library and the awaiting German soldier on the other side of the barrel fell. Immediate return fire began raining through the field as the Axis forces were caught off guard and scrambling to cover. With Bates near me and myself covered by three privates and two corporals we all began firing toward the enemy. Tunnel vision was not a question as I aimed at one soldier, took him out and fired toward the next. The Germans began to prone toward the ground and started to load their MG42's and began laying waste to our troops, one of their MG's took out seven men in our line. Under heavy fire we all ducked behind the wall, forcing us to cease fire until one of the other platoons could give us covering fire. Fear was of the essence as Axis troops ran across the fields attempting to hit us in our own trench. Three soldiers made it to my section killing the three privates to my left, leaving me defenseless on my left side. Bates and I began engaging in another hand-to-hand combat situation. The German on the left surprised me with a strike to my face causing me to fall back and drop my weapon. Incapable of reaching for my 1911 pistol I rushed the German bastard and began delivering strikes to his stomach, face, and throat. Bate's was busy with the other soldier but his battle ended rather quickly as he struck him and pushed him back far enough to pick up his rifle and empty 3 rounds to his chest. The German anticipated my strike and caught me by the arm, turned me around and forced me to the ground. He reached his arms around my throat and began to choke me. I thought I was done for as he locked my arms with his legs, to my saving Bate's came rushing in and smashed the butt end of his weapon across the Germans face and fired off two more rounds to his chest and upper head.

Catching my breath I collected my weapon, reloaded, gave Bate's a thankful nod and began returning fire once more. The battle was neck in neck as it seemed every time we dropped two of theirs they dropped three of us in return. I was next to one of the Browning machine guns and while I was reloading my emptied clip the Sergeant that was on it as well as his feeder took a couple of rounds to the head and chest causing them to fall off and leaving our weapon unmanned. I immediately jumped on it and ordered one of the corporals nearby to feed my rounds so they wouldn't get dirt inside the rounds and jam my weapon. I was unleashing hell until the sound of an aerial round cracked in the sky. Moments later an enemy mortar round slammed thirty yards to our left hitting a squad of soldiers behind the sandbags. All you could hear after were the gruesome yells of, "MEDIC, MEDIC, fuck, we're hit, MEDIC," and the medics began rushing over to perform aid on the wounded. I called our radio over and told him that he needed to get a hold of Captain Wright to get his men behind and flank their right side to neutralize the mortar teams. I ducked below the small hill of cover and grabbed the radio soldier by the arm and shouted, "Get Captain Wright on the phone and tell him we are pinned down by mortar fire, snipers, and uh… and kraut MG's, get their ass over there or we're fucked, go!" He radioed in, "Come in top three, I say again, come in Top three, thi…" Before he could get his sentence out a round pierced through his head and damaged the radio and began to smoke and within seconds communication was no longer feasible.

"FUCK, Bates, Bates, get over here now." Bates crawled below the hill to keep cover and rested right up on my right side, "Yes Walker?" He said as another mortar round had impacted on the open ground above us causing the dust and debris to fly over our heads. I pulled him in close and put his ear close to my mouth, "Radio's done fucked off, I need y'all to run a message to Wright and tell him that we are pinned down by enemy fire and mortars and we need them fuckers to flank their asses. We'll provide covering fire and you and… where's Ortiz? Ortiz! Ortiz!" Ortiz came up and laid in front of me, "Listen I need y'all to run with Bates and get the message to Captain Wright, flank those German bastards so mortars don't damn near take us out, roger? I shouted down the line, "EVERYONE LISTEN UP, ON MY ORDER, COVER FIRE SO THEY CAN RUN ACROSS THE FIELD, OK READY? ONE, TWO, THREE, COVERING

FIRE!!!" Thirty soldiers began firing upon the enemy and Bates and Ortiz hauled ass across the plain trying to reach Captain Wright's position. Bates had his pistol in hand and fired on the Germans coming out of their cover and dropped two before diving feet first into the lower levels of the hill. He began asking soldiers, "Where's CAP, WHERE'S the CAP?" They pointed him down the line and they made their way over until they found him ducked behind cover. "SIR, SIR, Message from Sergeant Walker." Wright fired off a clip of his M1 Garand and ducked once more, "What the fuck is it, can't he see we're not exactly at cease fire yet?"

Bates noticed a German sprinting toward their position and rose, fired two shots and killed him, "Sir they are taking heavy mortar rounds and can't move with them MG's firing, he said we need you to flank the right side and take them out or their fucked." He looked for his Sergeant that had been leading attacks with his squad.

Once found, Wright said, "Sarge, Sergeant, come here now, hurry." He met Wright at his position and the cap said, "Listen, Walker and Sorensen's platoon are pinned down by Kruat mortars and machine guns, take your squad and shut them fuckers up, we'll provide cover fire, find cover and throw some grenades and blast their asses, do you copy?" He acknowledged and was ready for the Captain's orders, "Alright on my go, Ready? GO!!!" We all stood up and began to fire until our magazines were empty. The Sergeant and his team sprinted toward the back where they were once again pinned down by the Italians protecting their mortar squads. The Sarge said, "Grenades ready, Pull pins, hold... hold, NOW!" They threw nearly seven grenades over and with it the soldiers were KIA. This gave us a fighting chance as we were now firing from in front of them and behind. We cleared the ambushing battalion with many casualties on both sides, we thought we were clear until we heard the roars coming from the ground as German Panzer II tanks were rolling our way, from the looks of it they had three tanks at the ready. Captain Williams noticed the approaching tanks and began yelling, "TANKS, TANKS, GET THE BAZOOKAS UP HERE NOW, LET'S GO!!!"

The four men carrying the weapons sprinted toward the tanks and took a kneeling position ready to fire. The first tank's machine gun began firing and hit the soldier on the left and caused him to drop the

Bazooka. The other privates began firing on the tanks, two of the three rounds had made their way through the first tank causing it to become disabled and left the Germans inside no choice but to exit. Once the hatch opened and they began running, all soldiers directed their fire toward them and killed them instantly. "One down, two to go," Said Captain Williams as he ordered the soldiers to reload. The second Panzer shot a round of his twenty-millimeter cannon and when it struck behind us it blasted its way through the sandbag barricade and killed four soldiers covering behind it. Captain Williams shouted, "Make ready, aim... FIRE!" Three more rounds of the explosive weapons raced toward the second Panzer, an immediate disable of the tank gave us enough time to head toward more cover. We had this in the bag, at least that's what we thought until the final tank shot of his cannon, the round directed right toward our Bazookas. They were hit and killed instantly and now left us with nothing to stop the final tank.

Low on grenades and heavy anti-tank equipment we thought we were out of luck. Bates noticed the Bazooka that was lying on the ground by the first soldier who got hit and said, "Sarge look, I can get it and get behind it to pierce the weak armor in the back of the tank, cover me?" I said ok and told him, "Alright but you keep your ass down, move quick, I am not really wanting to give your mom a flag alright? On me we'll cover fire and you'll move, no dumb shit please, alright here we go. MOVE!!!" He sprinted toward the gun, picked it up and began to maneuver behind the final panzer. He lined up his shot and released the round toward the back end of the tank. It went through and disabled the final tank. He quickly climbed up and opened the hatch releasing a grenade right on top of those bastards. Bates shut the hatch and jumped off before the explosion could cause self-injury. Captain Wright ordered us to push forward and engage the remaining forces. Nearly two hundred yards out was a German/Italian camp that had soldiers ready to defend. We posted up on the outside of the barricaded wall and got a quick accountability of ammo.

This was familiar to us as it was another trench like system that they had created. Captain Williams said, "Alright this is it boys, we take here and we may just defend Tunisia. Wright you go left, Lt, you take Sergeant Walker and your men and push right, meet in the middle of the camp when clear, roger?" We loaded our weapons and pushed in immediately taking fire. Lt and I were behind four other soldiers and began tossing grenades to neutralize the front men with MG's. We hit them hard and fast and before we knew it the Axis forces began retreating. Loyal soldiers kept firing our way but were killed soon after. With the Germans on the run we collected the surrendering soldiers and gathered them in the center of the camp. The Captains gave us orders to watch them and make sure they didn't try anything. Captain Wright approached one of the German soldiers and forced him to the ground asking, "Where's your commander's office?" Hoping that there would be sensitive information we could use to anticipate the Germans next move. Right after he asked this question the other soldiers behind him pulled out a grenade and ignited it. I raised my rifle and shot him in the chest before he could throw it. The grenade fell to the ground and was seconds from going off. I shouted, "GRENADE, GET DOWN!" I pushed Bates out of the way and without thinking tried to throw myself on top. Unfortunately, I was pushed out of the way by Captain Wright and he fell on top of the explosive. "BOOM!!!" Just like that we lost Captain Wright and the explosive caused us to fall to the ground to take cover. Once the smoke settled Williams asked, "Is everyone ok, and is anyone hit?" The smoke disappeared leaving behind the sight of Wright's body and Williams went over to check on him. "WRIGHT, WRIGHT, ERNEST!" He was KIA and we couldn't believe it. Captain Williams grabbed his dog tags and any letters or sensitive information and got up off his knee and said, "I'm sorry Ernest. Alright, get the POW's on the trucks and round up weapons and tags, gentlemen congratulations, we just won Tunisia." A couple weeks later we were headed home back to Fort Bliss. With nearly six hundred Americans

KIA, almost nine hundred wounded, and an estimated forty one Americans MIA (Missing in Action) the Axis forces suffered great loss and were defeated in Africa. Due to their loss and from the help of the Allied forces, France had joined the Allied forces and agreed to help in the fight against Hitler and Germany.

CHAPTER 9

HOME BUT NOT HOME

The next few weeks would fall silent as we were not preparing for another advancement. Within the last couple of weeks, Bates would join me at the rank of Sergeant and help me lead the soldiers in our platoon of the 509th Airborne Division. We strived to make the lieutenant's job easy and maintain focus on helping the captain and company commander. We formed the platoon outside the barracks as we were about to receive our next leave. This would be our longest leave since graduating from basic training. The company commander gave his speech, "Gentlemen, congratulations on your victory over the German and Italian bastards; your win dictates the strength the United States shows over its enemies, for we are no short of immortal. As I see it, you have all earned some time to go home and see your loved ones; use it; use this time to be normal again; use this time to be with those you have fought so hard to protect. Keep in mind, gentlemen, the wars are not over, so enjoy it but come back ready to destroy those fuckers, Roger?" We all responded simultaneously, "YES, SIR!" Days later, I was back in Georgetown, Texas, waiting for Thomas to arrive at the house; he would be there hours later.

When he returned, we didn't say any words, and yet, with the tightness in our hug, you could write a book with the thousands of words we were thinking. I sarcastically said, "Hey moron, are you hurt, huh? You hit anywhere?" and wrapped him in a hug again. He responded, "Nah, I'm good, Jay, I'm good, made it back like I promised, right?" We spent the rest of the night sharing our experience in Africa, and we must have stayed up for damn near ten hours just drinking and eating and enjoying each other's company once more as we did as kids. We spent most of our days working on the house and drinking cheap whiskey left at the house before we left for basic training. We grew tiresome from fixing the house and keeping ourselves busy, and when nighttime rolled around, sleeping was no easy task. Often, I would wake from a night terror that entered my mind as I could only dream about poor Durrago taking that hit. And it would only be when he fell and impacted the ground that I would wake up in accelerated breaths. Although this dream was haunting, it would never be the same dream each night as I would dream about the MG's that slaughtered men, young boys at that, boys who had nothing but the rest of their lives to live and with one small caliber of hot brass it was ceased from their grasp. They had no choice but to fight a war they couldn't confidently say they knew what they were fighting for.

The next night would be the explosion of the grenade that killed Captain Wright, and when I fell to the ground, that's when I would awake and find myself on the floor, my heart beating three times as fast and checking my surroundings as if the Germans were in my room waiting for me. With each waking, Thomas would walk into my room, not saying anything because he already knew that it was my nightmares that caused a stir. He often walked in and said, "Was it Durrago again, Jay?" I would always respond with a sentence that came out of my mouth quickly due to my breathing complicating and heart rate pulsing through the roof, "It uh... It... yeah, it was him, sorry Tom, You do not need to worry, these damn dreams, I am not getting any sleep, go on back to bed, I'm uh... I'm fine, goodnight." With each new leave day, I would leave the house and explore the town. Civilian life was almost suspicious to me now with everything that happened. The towns were quiet, lacking the sound of which I became all too familiar. Trucks would backfire, and I would duck for cover because I thought a German sniper was aiming at my position and trying to take me out. I was paranoid, not capable of calming myself down enough to be a part of society again as a normal person and not as a use of destruction paid by the government.

The next couple of nights, I would lose sleep, not due to the nightmares but because Tom found the opportunity to bring home some town girls. They would leave in the morning, and when he came to breakfast, I'd have the same sarcastic question awaiting him as he poured his OJ," So Tommy Boy, you uh, you going to marry this broad? I think she's perfect for your ass?" He grabbed the toast from the middle of the table and threw it toward my head, and said, "Fuck off Jay; we go back in a couple of weeks; who knows when I'll get to talk to all the beautiful ladies again." Grabbing the freshly made jam that I picked up from the market and smearing it across the toast that now laid in my lap, I said to Thomas, "Thinking bout see Ma and Pa today at the cemetery if you want to join, good weather and isn't too long of a walk anyhow." He lowered the coffee cup from his lips and placed it on the table, "Think they'd recognize us anymore, you know as sons and not killers?" I began to spread jelly on another slice of toast as I looked back and forth to him," Their mom and dad, they'd forgive us for anything. Hey, remember when we got in dads truck with his keys and pretended to drive around, and you told me to start the engine, and I did because my dumbass decided to listen? When he found us crashed into the fence, I thought we were done, but he came over, put his hands on our shoulder and asked if we were ok." Looking forward with coffee in one hand and toast in the other, he chuckled and got up to finish getting ready.

I stared out the window briefly because I knew this sense of home and peace would soon be over as leave was weeks from over. Within the next few hours, I went to the shop for some tools because our sink was leaking, and I needed something to tighten the bolt. That's when I saw a beautiful young lady walking into the bakery next to the tool store. She was a blonde, tall lady wearing her church clothes, a light blue summer dress, and a daisy resting on her ear. She had on white high heels. I walked into the bakery and approached the counter, where she welcomed me with a bright and kind smile and said," Good morning; how may I help you?" Shocked by her beauty and kindness, I hesitated to answer, and that's when she looked at me with a confused look and said, "Is there anything I can help you find." Pulling back to reality, I answered," Oh, yes, ma'am. Can I uh... get a slice of rhubarb pie, please?" She began to grab the slice of pie, and that's when I noticed the daisy in her hair. I said, "Oh, that's a beautiful flower; it looks great on you." She pushed back her hair to reveal the remainder of the flower. "Oh, thank you. There's a little floral garden by the church, and I go with my mother after church to pick new flowers for the window by the door; it just stuck out, and I decided to wear it," she gave a small chuckle and handed me the bag with the pie in it. "That'll be twenty-two cents, please." I handed her the change and told her to have a good rest of the day. As I walked out of the bakery, I turned back to look at her again.

Turning her head up and looking at me with a smile, she asked, "Is there anything I can get you?" I hesitated and walked to the counter again. "Yes, ma'am, I never got your name; I'm James, uh, James Walker." She raised her hand to shake mine and said, "Well, James Walker, I'm Leah May, Leah for short, very nice to meet you." With all the confidence in the world, I finished shaking her hand and made up a quick lie to hopefully help me win a date with her. I gripped the bag harder and said, "You know Leah, I haven't been in town for a while now, and I reckon I need someone to show me the new additions to town, maybe... over dinner?" She smiled and laughed as she handed me the receipt and said, "Dinner, huh? I'd like that; how about Reggie's Bistro, just across the street there? Hear the steak is to die for." I jumped back from the surprise that she said yes and responded, "Um, yes, ma'am, that's perfect. Let's say eight o'clock, and I can pick you up." She agreed, and as I walked backward, I stumbled onto a display table and knocked over some banana bread samples stacked on top of each other. Oddly enough, I couldn't stop staring at her beauty and wasn't watching where I was going. She smiled and said it was ok and that she would get it after I couldn't replicate the display previously set up. I walked out the door to get one last glance and went home to prepare. Walking into the house, Thomas stopped me and asked, "What's got you all happy and jumpy?" I placed the pie on the counter and responded, "I met someone at the bakery today, and she's beautiful and has the smile of a thousand bright suns." He looked at me with absolute disbelief, and up until now, I have never really taken a liking to a serious relationship. He said, "Huh, didn't think you were the serious type, don't pull moron shit and fuck it up; who knows when your ass will get another chance." I told him not to worry because I planned the whole date. First, we'll go to Dinner, then maybe an evening matinee.

Dressed in my AGSU with my hair gelled back, I left for the bakery to pick up Leah May as she got off work. We left for Reggie's and sat at a low-lit table in the restaurant's corner. I began the date by asking her how she got into baking, and she told me that ever since she was a little girl, she would always find herself in the kitchen helping her mother bake goods for neighborhood parties that were thrown due to the close friendships that each neighbor had built. She also told me how much she loved to hike and paint. There was a brief conversation pause after the waiter poured our wine, and I couldn't help but stare at her beauty again. With a slight chuckle, she stared at me and said, "What? Why do you keep staring at me? Taking a minute before I answered her as she sipped on some wine, I told her, "Because I think you're the most beautiful woman I've ever seen." She looked down at the table with a slight grin and softly said I must have been lying. The server brought out our food, and we engaged in soft conversation for the remainder of the Dinner. When we finished our meals, we left the restaurant but were stopped by the rainfall. I acted quickly, using my jacket to protect us from the rain, and we ran to the truck as quickly as possible. Her hair was drenched in water, and my suit was as wet as a running river. We just laughed it off as we truly enjoyed each other's company. We decided to skip the movie, drive around town, and continue our conversation.

As the night ended, I dropped her off at her house, and as we were saying goodbye, an intense sense of emotion came over me, and I leaned in for a kiss. As I leaned in, she felt the same and met me in the middle. We softly kissed goodnight, and she said, "I hope this means you'll see me again." Smiling at her, I said, "Yes, ma'am, I'd like that," she stepped out of the truck and walked inside but not before turning my direction again and grinning at me again. We saw each other a couple more times after that and grew into our relationship. A few weeks later, I received a letter saying that we were to report back to Fort Bliss in a week. Knowing that our time was short, I tried to see her as much as possible. I invited Leah to a picnic lunch in the park to break the news to her gently. I pulled out the blanket and prepared the sandwiches I packed a few hours earlier. She said, "So what's the special occasion?" Nervously laughing, I poured her a glass of champagne and said, "Um... listen, Leah, I uh... I received a letter in the mail the other day, and uh, I've been called back to base." She stopped sipping her glass and began to frown downward, as this was the last thing she wanted to hear. She was upset, and I knew she would be, so I hesitated. I put my glass down and said, "Look, I like you, and I don't want to lose you, but I have to go, Leah, I have to go. I'll write to you every day, and I'll think about you every day, and I promise to come back to you, I promise." she fought back the tears that were forcing their way out of her eyes and said, "You better James Walker, You come back to me you hear?" The night before I had to meet at the bus stop, she spent the night, and it was a night of nothing but romance as we remained close and intimate with each other.

The next morning, at the bus stop, we said our last goodbyes, and I raced to a window seat to give her one last kiss outside the window. She handed me a book with a photo of her in her nicest dress with curled hair, and she was smiling and looked gorgeous; she wrote a little note that said, "Come back to me, J, I love you, come back." As the bus took off, we stared at each other until she was out of sight, and before I knew it, I was back on base at Fort Bliss and in the barracks, where we all gathered to catch up with each other. Our next mission brief would be hours later and what would soon be our most dangerous mission yet.

CHAPTER 10

ALL'S LOST IN NORMANDY

Allied forces assembled in the late month of December 1942 to come to an agreement that our focus would be concentrated on defeating the German forces before attempting to neutralize Japanese forces. The first attempt was a success due to the invasion of Allied troops in Africa, which eventually led to one of the most brutal and unsettling battles of the war, D-Day. This attack would be planned out thoroughly, and the objective was to hit the less defended areas to allow for optimal air cover and let our amphibious troops invade the shore with high numbers and force. The objective was to drop airborne troops by air and drive in troops by sea so they could initiate the grand invasion. The British Army wanted to take advantage of the heavy German forces fighting in Russia and assumed this would be a less vicious battle. However, German forces were strong in numbers on this day and proved to be a challenge. Eventually, this invasion would be known as Operation Overlord. Amphibious training for U.S. troops occurred from December 31 through January 2.

This training aimed to help soldiers learn attack techniques that would be used for the invasion itself. It is estimated that nearly sixteen thousand troops were involved in this exercise. We knew we needed an efficient way of transporting troops onto the shore of Normandy. So it was found that Andrew Higgins, a New Orleans boat builder and inventor who had created the LCPV amphibious vehicle, was the brains behind transportation. This boat worked so the driver would be posted at the back, and the troops would be loaded in the open space with the doors opening at the front. Higgins would create nearly twenty thousand LCPVs during this time, but the numbers returned in good condition were less than a hundred. After we awoke in the morning, we conducted hygiene, ate chow and fell into formation for our daily brief. However, this brief would be devastating to most that our company commander would tell us that the inevitable mission would become a reality. Colonel Morris began his brief, "Good morning, gentlemen; today, we received information that Operation Overlord is a go. Training will begin immediately, and we must perfect our jumps, as this mission is paramount. The Germans will see death as they have so graciously given unto us. Army tops have given a plan to infiltrate Normandy and plan to do so by air and by sea.

Gentlemen, I cannot express enough the dangers that will follow, so stay focused, stay strong, and don't fuck around. Many troops will drop, but many of those soldiers will not return; how seriously you take this mission will determine which category you fall in. Good luck, gentlemen; I will let you know when we fly out as soon as I hear a word from higher-ups, and that is all dismissed." The only thing going through my mind right now was that I might not return to Leah and would have let her down. Captain Williams had us jumping day after day until the night we were supposedly supposed to drop into Normandy. The night came to be, and we all had our packs tight and ready to go. We received our last set of mail for a while. I heard the soldier passing out the mail calling my name, "WALKER... Walker?" I put my rucksack down, "Over here." I had 2 letters; one was from Tom, and the other was from Leah May. Tom's letter read, "Dear Jay, just got our mission brief, and it looks like we are hitting the beach first thing in the morning. If I don't make it back, please know that I love you, and Leah is way too good-looking to be with a sumbitch like you. Stay safe, Jay. Love Tom." Laughing at the joke about Leah, knowing he was probably right, I couldn't help but think, what would happen if Thomas died tomorrow? I'd lose a lot more than my sanity; I'd be losing the last family member in my life. I opened the letter from Leah, which read, "Dear James, I hate that you're not here; I constantly worry about you being over there. Remember you made a promise: be smart over there, and I'll be thinking of you here and writing to you every day; with love, Leah. I always felt better when I received letters from her, and it would often calm me down and put me in a place of euphoria, knowing that someone was waiting for me when I got home.

The supply soldiers began passing out our issues of MRE cans and cigarettes. Most soldiers finished the two packs they received within days. I was not much of a smoker, but due to the high stress and the need to calm myself down, I would usually smoke two or three every other day. Everyone was on the runway ready to take off, at least until we got word from President Dwight D. Eisenhower that the attack would be postponed twenty-four hours due to poor weather conditions over the English Channel. Meteorologists discovered that weather conditions would have a clear window on June 6, 1944, the day of the invasion. The night before we flew out, we took advantage of the free time, and many soldiers used it to write letters back home; many fellow Christians gathered with the Chaplain to pray and, for some, ask for forgiveness. I stayed with the Chaplain for a few minutes, said my prayers and went back to our barracks to write back to Leah, "Dear Leah, tomorrow is the big day; I just want to let you know that if I don't make it back I enjoyed every minute that I have spent with you, you are the most amazing women I have had the pleasure of knowing. If I am to go out, then you best believe I'm taking Krauts with me; I am not going out empty- handed. If I die, then please don't wait for me; meet someone and build the life you deserve. Expect my next letter within a few days, of course, if you do you will know that I survived and I am safe-keeping in France. With all my love and pledge to keep my promise, James. P.S.

We had Rhubarb pie for dessert; it doesn't come close to the pie at the bakery." President Eisenhower came over the radio waves and gave his famous "The Great Endeavor" speech, and the mission was a go. The next day, we loaded into the C-17 and were in the air within the next couple of hours. Captain Williams gave our platoon a mission brief on the plane before we jumped. "Alright, gentlemen, this is it, this is where we destroy those kraut fuckers," as he was giving the speech, we were minutes out of the landing zone, and we began taking fire from the German ground units. Many troops flying near us began jumping prematurely due to the planes losing their engines from the rounds impacting. Captain Williams stood up and prepared us to jump, "ON YOUR FEET, CHECK PACKS, and STANDBY." We numbered off, were cleared for drop and began exiting the plane. The line moved quickly, and before I knew it, I was following Bates out with Williams behind me. The skies looked like fireworks lighting up the sky, and tracer rounds from German MG42s flew by each soldier. I feared I would be one of the unlucky sumbitches to take three to the chest, but I felt a sense of God with me again and hoped he'd get me through one last drop.

Wind speeds picked up due to the weather, and this caused soldiers to be thrown off point and an oddly large gust of wind came by and hit Captain Williams. Such force from this wind pushed his body toward me, and we collided. The force of his body hitting mine nearly broke my left arm, and the air from my lungs was absent for seconds. He got back on course, and we were closing in on the landing zone.

The sights were horrific as I could see soldiers getting hit by ground units, causing them to be KIA in the skies before they even landed. Tragedy struck as I looked to my left to check on Captain Williams. I could see the tracer rounds closing in on his position. Unlucky, and as fate would have it, he was struck in the chest, and the remainder of the rounds tore his chute. He was killed, and with a torn chute, he fell to the ground as fast as a meteor entering the atmosphere. Captain Williams had died in the air and impacted the ground so hard that you could barely recognize the body upon landing. As he fell, I screamed, "CAPTAIN, CAPTAIN, and NO!!! SON OF A BITCH, WILLIAMS!!!" My heart was

racing, and I thought I would be next. I could see the rounds heading in my direction as if I were the only soldier left in the sky. Almost as if they were targeting me directly, I steered my chute left and right, trying to avoid the rounds, but I was unsuccessful as a round hit the center of my chute.

Luckily, it wasn't bad enough that I fell at rapid speed. I hit the ground hard, almost breaking my legs on impact. The pain from the forceful landing put me down for a couple of minutes after landing, but Bates rushed over to my position and said, "Walker, Sarge, you Ok? You hit? Where's the Cap? Did he land with you?" Grunting with the pain, I gave him the tragic news, "He, ugh, he didn't make it, sumbitches caught him out of the sky. Caps dead, where's the Lt?" Tilting his head down, he told me Lt Sorensen's location, and I gave him the whereabouts of Captain Williams's location, "Sir, Captain didn't make it, they shot him out of the sky, and he fell, I don't know where he's at now.

What's our next set of orders? You got command now,

Sir." Lt Sorensen couldn't believe the news and shouted, "GOD DAMMIT, he had our next objective ready, but he also had the map for our rendezvous point; Ok Ok,, round everyone up, Bates gets me accountability, Sarge, you get weapons and ammo check and have everyone ready to move in two minutes, we are moving to objective, go." With our platoon ready to move out and their M1s and Thompson Sub Machine guns ready, we formed a line and made our way to our next point.

Before we left, Lt announced to the rest of the platoon, "Alright, gentlemen, Captain Williams is dead; we move quick and quiet; nobody fires until I give the order, Roger? I got command now, gentlemen; listen to me and help me out, and we may survive this; alright, let's move." Once the Allies landed, their objective was to take Carentan, France. The idea was to gain a location or a defense line for the remaining invading troops. They would have had a better chance of pushing the German forces back if they had taken this town. Lt. Sorensen sent a runner to get information from the first attacking platoon. Here, we would meet another Captain who would eventually merge our platoons and take the lead of the troops. The runner came back with shortness of breath and told Lt, " Sir, I gave located the other platoon. I asked for their commander and they have a Captain named Gerald Jacoway. He has advised me to authorize our platoon to meet at Carentan. He wants us to move our ass as they get ready to invade the town." Lt began asking him for details of the German forces, asking, "Roger, how many troops are in the town, and what are we going up against? Weapons and any artillery or vehicles?" The runner took a sip from his canteen and caught his breath, "Um, he said, uh, at least fifty to a hundred troops, uh snipers appear to have taken cover in a buildings like second floor maybe, and they have at least three mg's from what the Cap could see, no artillery and only transport vehicles sir." Lt gave the next set of commands, "Alright, gentlemen, move quickly; we join the other platoon; when we arrive, we will take orders from Captain Jacoway, and we will help take Carentan, Roger?" We all responded, "Yes, Sir," and began pushing that way.

A couple of hours later, we arrived at Carentan and met with the other platoon. Staying low and behind the towns tree line, Lt Sorensen began asking for Captain Jacoway. Soldiers pointed him toward the Captain's position, and when they met up, Lt asked the officer, "Captain Jacoway? Is that you?" He responded, "That's me; about time you all showed up, thought we'd have to go in alone, alright look, from what we can see, they are mainly posted up in the center of town; we hit hard and fast and use the buildings for cover, I want four soldiers pushing houses at all times. Here take this radio; it's already on the correct channel; call in if shit hits the fan; on my go, we push. I want grenades ready for the M.G.'s. Who's second in charge for you?" Lt turned back to find me and pointed my way; not sure of his intentions, I began moving forward, and that's when Lt filled me in on the plan; he said, "Alright, Walker, Cap said we push in three lanes, you, Bates, and grab ten more guys and you push left hitting these four building here, use them for cover, put your radio on the channel to and when Cap gives the signal we move, roger?" I nodded and got my men ready. With three separate squads at the ready, we all anticipated the Captain's signal.

Captain Jacoway raised his fist, peeking at the wall, and when the Germans turned their back, he slammed it down, giving the signal to push. I sent the first shot down range, hitting the first German in the back and watching the blood exit from his chest; after that, all soldiers began firing. The German forces took cover behind brick walls and started to return fire our way. I yelled to my squad, "MOVE, MOVE, PUSH LEFT, CLEAR THE FIRST BUILDING, GO!"

There was an ally connecting to the first building, and we posted up at the first door; I signaled the breacher, and he kicked it in. I went in first, and three soldiers were running for their weapons; I fired at the first German by the table and shot two rounds. The force of the rounds caused him to fall over the table and land on the floor. The two soldiers who entered after me fired at the remaining Germans. One received a chest wound, and the other had a bullet pierced through his head, killing him instantly. We stayed quiet and lined up on the stairs. We softly walked up with weapons aiming at the next door. I raised my fist to halt everyone and told Bates to kick the next door. Once open, we began to receive heavy fire from Germans who had posted up behind a couch. Bates rushed back to his position, and I ordered one of the soldiers to prepare a grenade. They passed it forward, and I pulled the pin and threw it into the room. I yelled, "GRENADE, COVER!" it went off, and we could hear agonizing screams coming through the door, and we rushed in to finish off the survivors. Once the house was clear, I radioed to the Lt, "Lt, and first house clear, moving to the second house now, one friendly KIA, over." Lt Sorensen came back over the radio with sounds of gunfire and with him shouting due to the noise, "ROGER, KEEP PUSHING, CLEAR THE REST OF THE HOUSES AND MEET CENTER TOWN,

OVER?" We acknowledged his command and moved toward the second building. One posted at the next door, and we repeated our forcible entry at the previous house. I ordered the third man in line to breach and couldn't anticipate what would come through the door soon after.

Once he breached the door, he was met by heavy fire from the inside, and the German soldiers covered by furniture fired their Mp40s through the door and hit the poor bastard through the chest and head, and he immediately hit the ground. We waited for their reload, and Bates and I pushed through the door, firing off rounds toward the enemy. Bates shot the soldier to the right as I aimed my rifle at the one on the left; he saw his man die and immediately threw his weapon to the ground and surrendered. One of my soldiers approached the German and struck him across the face with his weapon, forcing him to the ground. He was held as a POW, and we pushed up the narrow stairwell to the left. I ordered my soldiers to watch the Germans as Bates and I walked up to check the room. I entered through the door only to find a woman and her daughter tucked behind the bed. The mother raised her hands and, in German, said, "Nicht schießen, nicht schießen," which translates to don't shoot. I lowered my weapon and told Bates, "Get them to safety and meet me at the next house." I had one of my German translators speak with the women. He asked her, "Gibt es noch mehr deutsche Soldaten? Was ist mit den anderen Häusern?" This translates to Are there any more German soldiers? What about the other houses?

She nodded to tell us not in this house. She told him that they had about ten more soldiers in the other houses with larger weapons and that they were holding two kids in the last house.

The German translator took the lady and the girl and offered to watch over them while Bates, the other soldiers, and I cleared the remaining houses. I gave orders to the three soldiers in the back to clear the third house while Bates and I took charge of the final house with the kids being held hostage. They entered the third house and began firing on the Germans. I looked at Bates and said, "When we breach, no explosives, we fire one bullet at a time, and the kids don't get hurt, Roger?" He nodded, and we were ready to breach. I could hear the Germans moving toward the door, and they opened it, and two German soldiers walked out. We knew the last remaining German was heading upstairs to lock the kids in a room, and the two were coming out to help with the other houses. We lowered our weapons and pulled out our knives to keep this quiet. When the two Germans walked out, Bates took the front man, and I took the German behind him. Bates covered his mouth and reached over and stabbed him in the chest, and I covered the back German's mouth and sliced his throat. Afterward, we ran up the stairs and kicked open the door to find that the German had the eldest kid at gunpoint, using him as a sense of protection. He began shouting, "Senken Sie Ihre Waffen, sonst stirbt das Kind! Lass sie jetzt fallen!" We knew we had to act fast. I looked at Bates and whispered, "Throw your weapon to the ground, and when he's distracted, I'll take the shot." I looked back at the German and tried to calm him down, saying, "Woah, woah, Ok, we're dropping them, easy, easy." I gave the signal, and Bates threw his rifle.

Once the German looked his way, I raised my rifle and aimed toward his head, releasing a round of his directions and killing him on the spot. The little kid ran to me and hugged me, and the little girl with him stayed covered under the table in the back of the room. We gently told her to come over, and when she came, we took them downstairs, and I had a soldier bring them to safety. I radioed the Lt again, "Sir, all houses clear, and four civilians found and brought to safety; what's your position over?" He responded, "ROGER, MEET AT CENTER HOUSE AND JOIN US BEHIND THE WALL, TAKING HEAVY WEAPONS FIRE AND

NEED ASSISTANCE, HOW COPY?" I acknowledged and ordered my men to rendezvous at the Lt's position. We pushed around the corner and met the Lt with the M.G. rounds raging against the wall. We took cover, and I crouched my way over to Sorensen, who was trying to make contact with Captain Jacoway, who had not yet radioed in.

Minutes later, they rushed around the corner, and Captain Jacoway was firing his rifle toward the M.G. until he took cover next to us. He devised an idea to lay down cover fire while we flanked them on the left side. Lt Sorensen prepared a smoke grenade and told us that when he threw the grenade, and the smoke blocked their sight that would be my time to make a move. I pulled Bates over to me, forcing his ear to my mouth and yelled, "OKOKOK, BATES, WHEN THE LT THROWS SMOKE, WE FLANK LEFT AND GET THESE BASTARDS, PASS IT DOWN AND BE READY TO MOVE

ON MY SIGNAL!" Captain Jacoway gave a three-second count, and Lt threw the smoke over. It went off, and the M.G. fire ceased; we ran to the left and maneuvered behind the building. We could see the Germans frantically moving to reload the MG's, and that's when we started firing on them. We took out four Germans, and the other three remaining surrendered. We took them as POWs, and it was at this point that we took the small town of Carentan.

CHAPTER 11

ADVANCING IN SAINT-LÔ

Allied troops began pushing into the town, and with them came more militarized vehicles, ammunition, food, and supplies. We took the night to celebrate our victory, and we must have gone through hundreds of bottles of whiskey and German wine left over from the Germans. Shipments of cigarettes swarmed the town, and each soldier must have had three packs on them; of course, with the high stress, they didn't last long. Captain Jacoway gave us a brief later that night and said, "Gentlemen, good and bad news. We have taken Carentan and pushed the Germans out of the town; the bad news is they have taken refuge in the nearby town of Saint-Lo. We move out tomorrow, and we fight them head-on. Word has it that the Kraut bastards have taken hostages and are murdering innocents and those who refuse to help Hitler in his cowardly advancements. Rest up, boys, enjoy the night; it might be our last for a while, dismissed." My squad and Bates took the second house on the left as our barracks, if you will. We prepared a meal and sat at the kitchen table, drinking liquor and smoking while we could. The night was quiet, and everything seemed calm for the time being. We tried our best to keep the conversations heavy, as we knew if we fell silent, the echoes of the war would crowd our minds, and we would be driven towards a dark place. Laughter filled the room as we discussed what we did before we joined.

A private named Juan Domego-Cortiz from Albuquerque, New Mexico, said he was a ranch hand before enlistment. He started off his story by saying," Back home, I remember trying to group up some of the cattle, and uh, I went over to round up this baby cow, and when I shoved him to get him to move toward the food, he got up all riled up and stomped on my fucking foot, damn near couldn't walk for three days. Fucker just walked away like nothing happened and started stuffing his face as if he didn't cripple my ass." We all began laughing, but soon after, those laughs turned to silent tears as he stated, "You know I'd rather be there getting stepped on every day than be here." With tears disabling his ability to speak coherently, he sobbed," Fuckers just shot him like they didn't care about who he was, who he could have been." He puffed his cigarette and said, "Poor bastard couldn't have been more than nineteen years old." He was referring to the soldier who breached the door and took the German rounds to the chest. He apologized for letting his emotions get the better of him. I reached over and grabbed his shoulder and said, "Listen, Cortiz, it was just fate; as fucked up as it is, I reckon he's in a better place; he gave his life for the better to stop these fuckers from killing innocents and hurting those who can't fight themselves; we push on, not forgetting the lost but making their deaths worth it; why don't you get some shut- eye, sleep it off, yeah?" He nodded and put out his cigarette. He stood up from the table, saying, "Yes, Sarge, I'll be ready in the morning, night fellas." Bates took the final sip of his liquor and turned in himself.

Once the boys slept, I polished off my drink and took a smoke break outside. When I was out there, I looked toward the night sky and saw bullets flying into the sky and loud bombardments in the distance. Steering away from my glare, I was immediately pulled into the conversation by Lt. Sorensen and Captain Jacoway; it was at an outside table in front of the house. He said, "Can't sleep, Walker? Come, sit; we're just chatting." I pulled up a nearby chair that had fallen from the second floor of the damaged building behind them and lit another cigarette. Captain Jacoway asked me, "How are the boys? Are you okay?" Taking a second to answer him, inhaling a puff of my cig, I slowly answered, "Oh, they're okay, just tired from today. I told them to get some sleep and be ready for tomorrow. Sir, do you think we'll push the Germans out?" With his leg crossed over the other and looking down, he stated, "Honestly, Walker, I don't know. I know we can't stop; we stop, and Germany wins, and I don't think I want that to happen. Each day is new; we'll take it one fight at a time. Hey Walker, how's your brother doing?" I couldn't believe I hadn't thought about Thomas once, with everything that happened, especially considering that he had rushed to the shores the same day. I looked at the ground, dusting off my boots, and said, "I don't know, Sir, I haven't thought about it too much today. God, I hope he's okay. I hoped for a letter, but I'm unsure if he's still alive." Captain Jacoway told me they had taken the beach last he heard, and the casualties hadn't been announced yet. He also claimed that the casualties were assumed to be high. The worry in my heart only grew larger as I thought the worst for Tom, and maybe this could mean that I would be the last living family member. I said goodnight to the officers and went inside, where I had readied a couch. I took some pieces of paper I had stowed away in my pouch and wrote to Tom, hoping to get an answer within the next few days.

The next morning, we all woke up and formed outside, where Captain Jacoway was ready to give his final brief before we rucked to Saint-Lo. "Alright, gentleman, today we save the town of Saint-Lo. Understand that this is a rescue mission if you will; Germans are terrorizing the innocents of the town and killing women and children for the hell of it. We will strike them and take out any threat. Our mission is to save them and rid the town of Germans. We move out in ten." I had my men refill ammo, gather food and cigs, and top off their canteens. We marched out and headed toward the town. We walked into the front of the town, and it was quiet—almost too quiet for comfort. Captain Jacoway had the men take a knee behind cover. He had sent one soldier to survey the first couple of buildings to see if he could see any sign of German troops. The Captain was uneasy that we didn't see any defense troops awaiting our arrival, considering we had received word that troops were flooding the town the night before. The soldier looked through the buildings and couldn't see any sign of Kraut. He turned our way and gave the thumbs up, and Captain Jacoway signaled that we should move forward.

Before we made it three feet, the cracking of a sniper round sounded through the sky, and the surveying soldier was brought to a halt. He looked down at his uniform to see a blood trail running down his jacket. He looked at us and fell to his knees and then laid flat on the ground as the round had stopped his heart, and he had died right before our eyes.

We were told to get cover, and before we knew it, we were bombarded by enemy fire. The raging rounds heading our direction had been successful, as they had killed seven more soldiers who were not quick enough to get cover. Surprised and caught off guard, I fell to my ass and started hurrying back to take cover behind the town wall. Two privates had prepared smoke grenades and tossed them to halt the fire so we could enter the town. We moved quickly and split between both sides. I left with Lt. Sorensen, and Bates went with the Captain and took the right. As I was running, I fired my weapon in their direction to provide extra cover for my men to take safety in the buildings. We had a sniper with us, and I ordered him to take the third floor and scope out the sniper. With him concealed, we could not maneuver in the middle of the town and hit our open allies. While he was taking the position, we entered another firefight with German troops across the street. We were off to a good start, killing a few Germans every few seconds. We were about to move up when they mounted two MG's on a crate nearby and fired in our direction. Two soldiers peaked above the wall that was providing us cover, and the Mg hit them with four rounds each, killing them and shortening our count. They pushed us back and caused us to be stopped in our tracks. I turned around to yell at the sniper on the third floor to take out the guns before they killed us all. I yelled, "HEY, TWO GUNNERS, TAKE THEM OUT SO WE...!" I was interrupted because the enemy sniper had spotted him and hit the perfect shot. His weapon dropped from the building, and his body fell out of the window, causing him to land on the ground behind us. I grabbed the Lt by the arm and pulled him in to explain my next plan: "SIR, I'M PUSHING TO THE NEXT BUILDING, and AND I'M GOING TO CALL FOR A FRIENDLY AIR ATTACK AND COVER MY ASS."

All the soldiers peaked at once and provided covering fire, and I sprinted to the next building. Halfway there, I was struck by an enemy rocket that had hit the ground next to me. The force of the blast caused me to fly through the building's window, which slammed me to the ground. Shrapnel lodged in my leg and was about six inches long. It pierced through my thigh, and I began to scream in agonizing pain. Lt shouted my way, "Walker! Walker! What's your status? Did you make it to the building?" Holding my breath, I finally grew the strength to answer, "I'm hit; rockets on the second floor next building; put some fire on their ass now." Lt strapped a rocket launcher attachment to his rifle and immediately hit the building.

The shot was successful, as once the building was hit, it threw the Germans out of the building, and the debris struck the M.G. gunners and allowed us to take shots. They were killed, and the squad was pushed up, and Lt. ordered a medic to my position. He gave me his belt as something to hold onto while he pulled out the shrapnel. Lt stated, "Walker, bite down on this, on three, okay, one, two." He pulled it out, and I let out a painful scream. The bleeding was aggressive, and I was losing blood in seconds. The medic applied a tourniquet and some morphine, and I was back in the fight. I asked the Lt, "Sir, what's the Captain's position? He responded, "They're good. They joined us in the middle of town, but we couldn't get past the troops holding the tower; we suspected the sniper to be in the bell tower. He's taking us out like moths to a flame. We need to order the air strike. Can you call? I'll clear our troops and be ready to take cover." My ability to walk was hindered, and my movement was slow, but I couldn't let that stop me from leading my men to victory. I used the table next to me to leverage myself up, pulled out my radio, and called, "Air one, Air one, this ground two, over? “A few seconds later, they responded, "Ground two, this is Air one; go ahead." I yelled to Lt. Sorensen, "SIR, BE READY TO TAKE COVER!"I radioed, "We have large enemy troops on our position, and we can't push through the town; we need air assistance on my smoke, over? “They said, "Good copy, Ground two, air strike five minutes out." I yelled at Bates to pop smoke on the tower and take cover.

We could hear the Douglas SBD Dauntless bombardment plane approaching, and Captain Jacoway yelled, "EVERYONE DOWN NOW!" The plane released two bombs toward the tower, and the sniper took one last shot, killing a soldier who wasn't behind cover enough. The first bomb hit the tower, causing it to fall, and the second bomb hit the surrounding area before it and neutralized German troops, laying fire on our position. The blast impacted a couple of buildings nearby, and we checked to see if there were any German survivors. Limping into the building, I walked through the house and eventually up the stairs. I breached the door only to find a horrific scene. A family of three civilians were killed in the bombing; all three had burn wounds, and the young child had missed her leg. The scene caused me to become nauseous, and I vomited going down the stairwell. Tears overflew me, and I had to step outside and move behind the building to gain my composure. Private CortIz came to check on me. "Sarge? Sarge? Are you good? “He put his hand on my back, and due to the intense emotion, I swatted it away and raised my rifle to his chest. “Get the fuck off me," I shouted. He raised both hands in the air with fear consuming his body. "HEY SARGE, EASY, ITS ME, ITS CORTIZ, DROP THE GUN!" Exhaling large

amounts of air and unable to lower my weapon, almost as if my body froze, I gained control of my body again. I apologized to Cortiz," I'm, I'm sorry, I'm sorry, just uh, I'm okay, I'm okay." I lowered my weapon, and we searched the other buildings. I soon realized that the air strike was an emotional decision that was not thought out.

With every building we swept, the sight of innocents overwhelmed us, as the sights were disturbing. With each victim, the rage inside us grew, and the desire to kill every German here was vastly increased. With the German tower being destroyed, it left the Germans vulnerable, and the way that we were clearing houses, they wouldn't be able to overwhelm our forces. We came across the town's bridge that overlapped the flowing stream through the town. We managed to push the Germans across to where they barricaded themselves in the houses and used damaged furniture as cover. Captain Jacoway ordered, "Push forward, get them out of those houses, and push through the back of town." My squad was the second team, and we began firing on my positions again. We used heavy rocket munitions to hit the first floors of the building to see if we could injure the Germans on the next floors. We were successful, and Bates used a rocket to hit the building to the left. He fired the round inside, and it caused the ceiling to collapse and the Germans to fall. When they landed, we shot and killed the four soldiers struggling to get back up. This battle was surely ours as Germans began retreating out of the town; we thought we had one until Bates pointed out past the weapons, smoke concealing the view behind it. A battalion of enemy soldiers approached our positions, and we were outnumbered three to one. Captain Jacoway told us to get back across the bridge and find cover.

My squad began to run until we were pinned down by heavy fire. The Captain's men gave us cover fire, but it was enough to get a few men across at once. I grabbed my last smoke grenade and tossed it before our cover. I ordered Bates and Cortiz to run like hell and get to the other side of the bridge, and I provided covering fire until they made it over safely. With Bates and Cortiz moving and the smoke clearing, I knew I had only a small window to join them. I could hear Captain Jacoway making the call, "Come in Air, Come in Air; this is Captain Jacoway; we have heavy enemy troops on our ass, and we need to clear them out; drop on my smoke." They responded, "Roger, sir, drop two minutes out, take cover." Jacoway threw the smoke as far as he could and it landed in front of the Kraut's position, and I began to sprint to the other side. Only ten yards from the bridge, I felt a sharp and hot pain enter the back of my right calf. I was slammed to the ground, only to find that a bullet had entered my leg. Seeing my collapse, Bates and Lt Sorensen ran to me to try and help me up and shouted, "WALKER, WALKER, COME ON MAN! GET UP GET UP." With the bombs approaching, I knew I wouldn't have time to crawl over, so I made the call, "Sir, Bates, they are going to take me." Bates looked at me and said, "Walker, I can't leave you, man. Can you walk? "Grunting and struggling to stand up, I fell back down. "LISTEN, I AIN'T GOIN ANYWHERE, GET THE HELL

OUT OF HERE, GO BATES, GO.!" Lt. Sorensen ordered Bates to move, shoving him away. "GO BATES, get out of here, go, I'll get..." Before he could finish speaking, two rounds from a luger pistol entered the Lt's back, and he stood there as he watched the blood drip from his chest. "L.T.!" I shouted as I watched him fall. Lt. Sorensen had been killed trying to save me. Bates made it to the other side of the bridge, but before he could run back, it collapsed.

A German commander came to me with a soldier at his side; in German, he said, "Take this one; we can use him for later, alive for now." The German soldier struck me in the face and knocked me out, and before I knew it, I was on a truck heading south of the town and had no clue of the prison to which I was going. The Germans could hear the plane approaching, so they got in their trucks and exited town. The bombs dropped and destroyed nearly half the town with them. Bates rushed to the Captain to inform him that I had been taken, Lt. Sorensen had been killed, and they were going to bring me to a POW camp. Captain Williams looked at Bates and asked him if I had any family members that he could inform of my current POW status. Bates told him that I had a brother in the 32nd Infantry Division and that he had stormed the beach on June 6th. He gave the order to Bates, " Alright Bates, I am sending you as a runner, find the commander and Thomas. Let them know we need their help finding Walker and any U.S POWs. Join the 32nd and meet us there."." Bates took the civilian car just before the town entrance and took Ortiz and Cortiz with him as they made their little squad. They proceeded toward Orbec,

France, where the 32nd had made way after they took the beach, and were told that this is where they made camp for the time being. They reached the position of the 32nd and found the commander. Bates saluted the Captain and said, "Sir, my name is Private Bates; I'm with the 509th Airborne Division." Soon to be known as Captain Nile McDonald, also known as "Mac," did not salute back but rather aggressively said, "I don't give two fucks who you are, why you are in my tent right now, Private, can't you see I'm a little busy here? "Stunned and hesitant to speak again, Bates responded, "Sir, I have a message from the commander of the 509th; it's regarding Sergeant Walker. Germans have captured his brother James Walker, and we are here to give the message to his brother." The Captain walked around the table and sat on the front side of his desk and said, "Well, you might be shit out of luck there, Private; Sergeant Walker has been captured by Germans as well; son of bitch might be KIA by now; God willing, that is."

They each looked at each other, wondering if I were to be found again and how they would give the news to me. Bates returned to the attention position and said they had a proposition to help. Bates stated, "Sir, the commander is requesting assistance from the 32nd to help save

American lives from a German camp. We believe them to be holding Sergeant Walker there, as well as other American soldiers and innocent civilians." Captain Mac looked at them, pulled out a cigarette, and lit it. He took an intense puff and softly said, "Look around, boys, not like we have too much to spare. They messed us up pretty well on the beach. Do you boys know what a man can do to another man? The horrible demons war can bring out of a man?Shit, you'll never forget. Forces were strong approaching the beach, and within minutes, most of them were killed before they hit land." Bates noticed that his hand was trembling as he raised the cigarette to his mouth; he could tell that he was traumatized and was not willing to fight another battle unless they had to. Bates said, "Sir, I know this isn't an emergency, but if there are Americans at that camp, then we owe it to them to try to save them.

Killing Americans is one thing, but for them to kill innocents, women, and children, well, then we need to stop them before it's too late." Captain McDonald agreed and had his men ready to move the next day. " Bates showed him a map and pointed at a location they believed I was being held. He said, "Okay, if we do this, then we do it right and hit them fuckers hard. It'll take every man to get in there, and our goal is the Walker brothers, civilians, and POWs if we can, Roger? We don't need another Normandy beach shitshow again." We agreed, and Cortiz asked the Captain, "Sir, what happened on the beach? "He could not answer and dove into a frozen state with the echoes of screams and soldiers begging for their lives. He snapped out of it and said, "It was horrible; they were prepared for our arrival. I remember Sergeant Walker standing beside me in the boat, ready to advance. How could we have known what awaited us?"

CHAPTER 12

THE RED OF NORMANDY

The waters were unstable, the boats were in unsettling conditions, and they were made for quantity, not quality. Rain came down on us like the low ocean levels needed to be filled. Now known as Sergeant Thomas, he was in the first wave of the formations and was approaching the shore quickly. He was next to Captain McDonald, who briefly described the mission and how they needed to complete it. With the roaring waters, it was near impossible to hear what he was saying; the Captain shouted, "Alright, gentlemen, this is it, this is where we stop these bastards, do you hear me? No longer will they rain terror on innocents, and our chance to stop Hitler is here. When we hit the beach, I want Rifles first; after them, I want Bangalore's ready. Do not take your weapons out until we are thirty seconds out from the beach. Two men together, so if someone drops, you can take ammo and supplies. Gentlemen, thank you for your service to a great cause and May God save you. Be ready to move; good luck, gentlemen." The driver of the boat shouted, "THREE MINUTES OUT, BE READY." Thomas took a look around at the soldiers around him.

Many became emotional, and he could see grown men crying and exhaling aggressively. They were scared, and for good reason. The German forces were prepared as they blocked the beach with Czech hedgehogs, which were concrete barriers to block any amphibious vehicles from coming on the beach. These did not provide much cover and left bodies vulnerable. Many American soldiers began praying, hoping that God would come through and allow them to see another day. One soldier next to Tom became nauseous and began vomiting near him when the vomit hit his right boot. Thomas gave him a slight grin and said it was OK. He put his hand on his shoulder and gave him a prayer: "Dear God in Heaven, let not this man fall in the rage of war; protect him and get him home safe; watch over him and keep him from harm; in the Lord's name we pray, Amen." Stay strong, he said, and ready his weapon. We were approaching the beach, and the driver gave us the final warning: "THIRTY SECONDS." Captain McDonald yelled, "Weapons ready." Every LCVP began taking fire, and all the soldiers ducked to remain behind cover. The LCPV docked about ten yards away from the shore. Most of the vehicle's hatches were dropped, and many of the soldiers within them didn't make it to land. Many were caught in the fire and killed instantly.

From the point of view of the driver, when approaching the beach, he saw nothing but German bullets hitting the vessels and the sounds of mortar fire cracking in the air. A Sergeant First Class had ordered the driver to drop the hatch, but he had refused, stating to the Sergeant, "I can't do that now, Sergeant; we are roughly in twelve-foot- deep waters; with your pack's weight, as soon as you touch the water, you'll drown." The Sergeant replied, "Fine, but hurry up and get us to shore." The driver hit shallower waters and then released the front hatch of the vehicle, and within seconds, the fire of the MG42s had raged through the soldiers of our vessel and killed about seventeen soldiers before they could react to anything around them. One soldier had lit a cigarette before the hatch dropped, and once it opened, a bullet entered through the head, and he fell to the ground instantly. Terrified and low on options, a few of the evading soldiers used the dead bodies as cover, even though the number of bullets impacting the bodies was not enough to protect them, and the situation turned fatal for most. Tom was toward the back of the vehicle and was spared enough time to evade the rounds.

Shortly after, Captain McDonald ordered the remaining men to climb over the vehicle and land in the water. He then grabbed Tom just before the German rounds made contact. Tom was forced off the boat so hard that he landed on his back, hitting the water so hard and knocking the wind out of him, which hindered his ability to swim toward the shore. Nearly drowning, he was able to pull himself up from the water, only to notice that the soldiers around him were taking heavy fire. As Thomas was walking up the beach, a mortar round had impacted only twenty feet in front of him. The force of the impact caused Tom to fall back, and the blast of the mortar hit three soldiers, causing their legs to tear off. One soldier's arm had flown above Thomas's head, and the unlucky soldier began screaming in agonizing pain. Medics could not do much for the dismembered limbs, so he gave the man morphine and moved on. Captain McDonald ordered the men to push forward. Thomas had taken the rifle out of its Cosmo line wrap, preventing it from water damage, and began firing back toward the Germans. With the weather conditions, rainfall, and fog, the Germans were well-concealed and accurate, and shooting proved difficult. The only way you'd be able to get a clean shot was by waiting for the MG42 fire to give sight to the German's position or moving closer to the enemy. Soldiers were posted on the Czech barriers, but many were left vulnerable due to the shallow cover. Captain McDonald had ordered three men near him to move to the next barrier as standing still would be a death sentence. They lowered their weapons so that they could sprint to the next barrier, and as they moved up, the MG fire rushed through their bodies, with one soldier falling victim to three rounds through the neck and face and the others taking fire to the lower abdomen and legs. They were instantly KIA.

Many men were witnesses to these gruesome deaths and became

emotional. Tom remembered looking at one specific soldier, who could not have been more than nineteen years old, shouting and begging for the Lord to help him. The boy began saying with sobbing tears, "Lord, please, I don't want to die, Lord, please help..." and before he could finish the sentence, a round hit him in the chest, and he fell without a sound to follow. A majority of the men didn't make it halfway to the beach. Thomas looked around and saw nothing but dead bodies and the horrific sight of men with open wounds and blood trails stretching a couple of feet in each direction. As Tom pushed to the next barrier, he noticed a young man whom a previous mortar round had hit; the wound was one that he would never forget. His stomach wound was a gaping hole, with his lower intestine peeking out of the wound. Tom applied pressure to the bullet wounds in his legs, and the medic applied morphine to keep him comfortable. The medic looked at Tom and said, "Listen, with his wounds, he won't make it another ten minutes, alright? We need to move on those who have a chance. I gave him morphine, and this should keep him comfortable, but move on." crying and begging Tom not to leave, he had no choice; he was too much in the open and was subjectable to German rounds. Turning his head to the Ground, he looked at the dying soldier and said, "Listen, I'm sorry, but there's nothing I can do, alright? I'm sorry, the chaplain is here; let him pray for you. I'm so sorry." He grabbed Tom's arm so tightly that when Thomas tried to move away, he was pulled back to the Ground and forced to stay at his side. He pleaded, "Please, no, please don't leave me; I'm begging you, please help me." Tom had to yank his arm free and crawl backward, staring at him as he slowly passed away. He rested upon a barrier, just staring at the soldier, completely blocking out the gun sound and mortar fire, and he took a second to feel the guilt that overwhelmed him for leaving his side. Tom came to and shouted, "Fuck, Cap, we aren't going to last much longer if we don't get to those wires. What's the plan?" He began to speak but was interrupted by a mortar round that landed a few yards away, killing three more men. Captain McDonald shouted, "Wait for the reload, and then everybody pushes up, ready, and go!" We all ran to the hill, overshadowed by barbed wire and covered by a low hill, where they could lay down and be concealed from enemy fire. Tom had taken cover and looked back toward the beach. Disturbing images were left in his head as all he could see were innocent men slaughtered in the heat of battle and medics trying drastically to pull the surviving wounded men to cover to perform aid.

During this time, it was nearly impossible to see the sand that overlapped the beach due to the number of bodies lying on top. He thought to himself, "How could they send us here? How could they allow for this many men to perish?" He turned back and crawled to Captain McDonald. "Bangalore's! Bangalore's! Let's go; get your ass up here," McDonald shouted. Two privates approached with the Bangalores, but one Soldier was caught in the crossfire and dropped the explosive on his way up . It fell, hit the sand, and was in direct fire from the enemy. "Give me covering fire, and I'll get it," Tom shouted as he sprinted toward the explosive. All the men stood up and unleashed fire upon the bunkers, causing a ceasefire of the MGs long enough for Tom to apprehend the Bangalore. He dove back down to the hill and prepared the explosive. He tossed it to the front of the barbed wire, and it was destroyed, giving them access to the frontlines of the German trenches. The smoke of the blast gave them a small window to run up the hill, where they would no longer be pinned down by heavy German weaponry. Soldiers who had packed TNT in their packs lit the attached fuse and took cover before the bunkers. Once the fire ceased, they tossed the TNT into the bunkers and took cover to avoid the blast. You could hear the Germans screaming, trying to grab the explosive in time, but it was too late. The TNT detonated and destroyed the bunker, killing the Germans inside.

This was the first of their victories, as they would no longer have to worry about the MG's laying waste to half the battalion. They used the destroyed bunkers as cover and began firing toward the Germans on the hilltop. Moments later, the next wave of LCVPs had docked on the shore, allowing more soldiers to join the fight. Tom's next objective was to take the top land where German forces were bunkered down and had trenches dug so they could move around behind cover. With allied forces replenished, they advanced toward the next trench. Captain McDonald had ordered his men to form and clear each bunker for the remaining Germans. McDonald told Tom, "Alright, take men inside and clear the bunker; head left through the trenches and meet at the middle ground outside; move, move." Tom swept left, where he was met by two Germans running out; he raised his Thompson submachine gun and fired rounds toward them. He struck one in the chest, and the other was hit in the helmet, causing it to fall off; his next set of rounds had entered the German head, dropping him to the ground. He looked behind him and said to the soldier, "Grenade, hurry, come on, come on." He pulled the pin and waited two seconds before throwing it inside the bunker. Once the blast went off, he could hear the yells of the Germans, and he advanced inside. The desks provided cover, and the men engaged in a close-quarter firefight.

Tom killed the Germans, crawling away after getting hit by the grenade and killing them on the spot. The three soldiers behind him fired toward the Germans, taking cover behind the desk, and killed them before they could run out. Tom said, "Clear, push to the next room, go." They turned left to move in, where three German soldiers met them. One had knocked the weapon out of the Allied soldier's hand and began striking him with his knife. He stabbed him in the chest and killed him, and that's when a friendly soldier pulled out his 1911 pistol and shot the German three times. The other soldiers had fired at the next two Germans, and they fell to the ground. Before they moved on, Thomas told his men to collect anything useful. They gathered maps and letters—anything that had a symbol of significance—just in case it had valuable information. They made it to the final room, where they found five Germans, including an officer, trying to surrender. Without hesitation, the Allied Soldiers shot and killed each of them, as the rage of the Germans killing their brothers had deemed them as ones who don't provide mercy. Thomas stated, "I don't care if these fuckers try to surrender; you kill every last one of them. Yes?" They all responded, "Yes, Sergeant." Thomas and his men left the cleared bunker and met with Captain McDonald, who had also lost a few men while clearing their bunker. He had told Thomas that his bunkers were clear and that they had collected sensitive items and maps and stowed them away in their pouches. Captain McDonald asked Thomas, "OK, so what are we looking at now?" Thomas had told him that the bunkers were clear, but they were still taking heavy fire from the MG's on top of the hill.

Our platoon was in a position where the beach overview stretched for miles, and we could see our ships in the distance. Captain McDonald saw the cannons from the ship and came up with the idea to call in a strike and hit the Germans on the hill. He pulled me close and said, "Tell your men to get cover and keep their heads down. I'm calling in a strike, and we will wipe the bastards out." Tom yelled to his men to get their heads down and stay low. Once in cover, Captain McDonald pulled out his radio and contacted the ship, "Arkansas one, Arkansas one; this is Ground; come in, over?" After a few seconds of silence, they answered back, "Ground, this is Arkansas; go ahead, over?" McDonald's tried to get the information to the ship, but it was cut off by MG rounds hitting the bunker that separated them from the Germans. Once the gunfire was silenced, Captain McDonald gave the coordinates to the ship: " Arkansas one, the heavy enemy fire is pinning us down and we are requesting assistance, over?" Allied soldiers tried to reduce the amount of fire heading their direction by engaging in battle once more but were unsuccessful as when one soldier overlooked the cover in front of him; he was met by a round that struck him in the neck. He fell to the ground and began suffocating. Pressure was applied, and medical treatment was started instantly. Treatment was of no use, and tragically, the soldier had died.

Arkansas one finally responded, "Roger ground, keep heads low and stay clear of blast, fire 20 seconds out, over?" Captain McDonald copied, and the cannons fired. With their heads tucked in the dirt, they could all see the mortar rounds fired from the ship, which could be seen flying through the sky. The rounds had struck the Ground in front of us, and the MG fire had ceased. All that could be heard were screams from the Germans and the sound of falling barricades. The dirt and smoke flew over their heads, turning the day into night for a few seconds. Once the dust cleared, they noticed a German soldier sliding down the hill in critical condition. Unfamiliar with the German language, nobody could understand what he was saying, but it was obvious that he was begging for medical aid. Thomas called for a medic to come and treat him, and as he was giving the German aid and Thomas was on his knees helping, the sound of a pistol cracking off caused Thomas to stand back with fear. Captain McDonald had killed the German with his pistol. Angered and confused, Thomas angrily yelled at McDonald, "What the hell are you doing? He was no longer a threat, you asshole; you can't just kill him, you piece of shit at least the Germans we killed in the bunker had weapons and were ready to fire again." Thomas grabbed him by the chest and began pushing him back. Captain McDonald then struck him across the left side of his jaw, and Thomas fell to the ground. "Don't you ever put your fucking hands on me again, that's your one warning Sergeant. Would he have let you live if the tables were turned? You'd be lying in the fucking dirt with a bullet in your head and a folded flag at the ready for your next of kin. They are here to fucking kill us, end of story. We don't take prisoners, Thomas; we terminate the enemy and move on. Do you understand me, Sergeant?" Ignoring the Captain and refusing to answer while the soldiers behind him helped him to his feet, the Captain asked once more with increased aggression, "Sergeant, I asked you a question. Do you understand?" Wiping the blood from his lips, Thomas responded angrily, "Yes sir, I understand." He ordered the men to gather their things and start moving again. They were closing in on victory; the only thing that stood in their way were the remaining trenches and bunkers holding German defenses. They all formed one large line and proceeded forward through the trenches.

Thomas had heard Germans on the other side of the trench and readied a grenade. He tossed it over and killed three Germans approaching the entrance of the final bunker. The Allied forces risked death as the only way through the bunker was the front entrance due to the barbed wire panning across the trench walls. Thomas said he would go in first and engage the enemy head-on while the other soldiers followed. Captain McDonald agreed, and they moved in. While Thomas was heading inside the trench, he accidentally slid his feet across the wired trap below and triggered an explosive that was a defense mechanism for the German's escape. The blast sent him forward and acted as a concussion. He lost his hearing, and shrapnel pierced his abdomen and entered his legs. The debris blocked the entrance, and the other soldiers could not escape. Captain McDonald shouted, "THOMAS! THOMAS! Can you hear me? THOMAS?" This would be the last time that Thomas would hear his voice. An officer of the German forces entered the bunker once all was cleared and looked at Thomas. He said, "Take him to the camp, make him suffer; he may have valuable information. Put him on the truck, and let's move out." Thomas was loaded on the truck, and the Germans retreated. McDonald and his men saw the Germans retreating and fired on their position; they were able to get the runners but were unsuccessful in saving Tom. "What now, sir?" said a soldier to the Captain. Captain McDonald said, "Set up camp, prepare the tents, and allow other waves to come in. We wait here for our next command. Thomas was taken, and we need to find out where to go." D-Day would end up being one of the most brutal and one of the highest casualty recordings of the war, with nearly four thousand Allied forces being KIA or MIA.

With victory as a result, it didn't feel that way with the number of losses the U.S. and Allied forces received. The camp was made, and only a day later, the runners from the 509th airborne division would show up with the news. Myself and Thomas were now known as POWs of the war and were unlikely to survive. All I could think about while I was being transferred to this camp was Leah. How could she know? Who would tell her? Unable to send letters and not likely having the chance to do so anymore only caused my heart to sink even more. I promised her I would return and keep it, no matter the cost. I may not be a bright man, but one thing is for sure: I was an honorable man who kept his word and always came through on his promises. I wouldn't know that Thomas was captured until I arrived at the camp. Only here would I find out that the worst of the war was only about to begin. Once Bates, Ortiz, and Cortiz arrived at the camp, the plan was made for the 509th and what remained of the 32nd to merge and prepare a rescue mission to save Thomas and I. Captain McDonald looked at the men and said, "Alright, we can't move out now. We are too low on munitions and supplies, and we won't be restocked for a couple of days; all we can do for now is pray and hope to God they survive until we get there. Once we are ready to move out, we will use our vehicles to get there and then proceed on foot. I'll meet with your commander... uh, Captain Jacoway, Right?" Bates responded, "Yes sir, of the 509th." Captain Mac said, "OK, I'll meet with him, and we can devise a plan there. Send one of your men back to deliver the news and expect our arrival in a couple of days, no more than a week. Dismissed, gentlemen." Bates and the two privates saluted the Officer, and Cortiz drove back to the camp in Carentan. Bates walked away and pulled out a cigarette to light. Ortiz asked Bates, "You think he'll make it that long?" Inhaling a puff and releasing the air, Bates said, "Walkers are a tough sumbitch. These Germans are tough too; we're going to make them pay, that I know for sure." Bates continued to smoke and look to the sky and said to me consciously, "Don't worry, Walker, we're coming for you."

CHAPTER 13

AUSCHWITZ

Captured and guarded in the truck, movement was quick and efficient as the Germans brought me toward the camp. I could hear guards yelling at POWs as they approached the drop-off location. My vision was impaired due to the black bag over my head and the beating I received every few minutes during the journey over here. My eyes were bruised, and my body had no feeling other than the aches soaring through my bones. I feared I wouldn't make it through the night, let alone the fact that I was a US soldier entering the enemy's concentration camp. All I could hear were the sounds of terrified tears and hesitant footsteps as the loud creak of the gate fence opened and guards moved prisoners forward. I was pushed forward by the buttstock of a German rifle and placed at the back of the line, heading into the camp. "Move, American," said the German, aiming his rifle at my back. They finally removed the bag from my head, and the brightness was overwhelming and disorienting. A German guard at the front wave waved his hand back and forth, telling the other prisoners to move forward. One lady was ordered to walk but hesitating, as when she looked in front of her, seeing the majority of the camp, she began to freeze in place. The guard grabbed her by the arm and said, "Let's go and move forward now." She fought him a little, trying to release his grip from her wrist, and when he slipped off, he raised his MP40 submachine gun toward her and yelled, "Move now, or I will shoot you. Go!" Overwhelming tears flowed through her eyes. She screamed and ran in the opposite direction, desperate to escape. The guard ordered the men to stay clear as he raised his weapon and fired a round in her direction. She made it about twenty yards before she was struck in the back and her sprint was halted.". She fell to the muddy ground and was killed right then and there. The guard steered his weapon towards us and stated, "Anyone else going to run? You will listen to my commands, or you will be shot! Now move it." We all stared at the corpse in the mud for a couple of seconds to show sympathy for the brave young lady and then shuffled our feet forward until we were through the gate.

Concerned that I would be made an easy target due to my damaged leg and uniform, I kept my head low and followed directions to the letter. Pushing forward, I was met with a German fist across the left side of my jaw that caused me to fall to the ground. "Get up, American filth; the commander is going to love you; get up and get your new uniform, now!" I was in line to get the prison uniform, a grey and black patterned shirt with matching pants and shoes. On the clothing were numbers stitched on the right side of the shirt; the numbers went from one to the hundreds. I grabbed my clothes and stripped down in front of everyone; Privacy was of no importance to the guards, and with a thousand eyes on me, I put on my shirt. I was no longer James Walker from Georgetown, Texas; I was prisoner number Eight Seven Nine. My belongings were seized, and just like that my tags and letters I had to Leah and Tom were taken, my rank and name were removed. I was pushed out of line and told to move on to where I would be held for God knows how long. At first, I thought, Okay, this isn't so bad; maybe I'm just restricted from my common rights and don't have many options to keep busy.

Unfortunately, my thoughts would be proven wrong as I passed the clothing line and turned the corner of the street. I could see in the distance German soldiers dumping bodies on the ground and being piled up like trash gathering on the street. Most of the bodies were stripped and beaten, and from the looks of the rib cages poking through the skin, these bodies were malnourished, almost like these poor souls hadn't eaten in weeks. " I was halted, pushed to the ground, and they slammed the gate doors shut behind me. I picked myself up and limped to the gate, where I rested. I rested my arms above my head and peeked through the camp to find my best chance of escape. Across the street, I could see the women locked away. Many of them were mothers desperately holding onto their children, or in some cases, children. I particularly noticed one mother and her daughter. The daughter could not have been more than twelve years old. I could tell because of her young features such as height, colored hair, and unwrinkled facial structures. A gentleman approached me on my left side and rested up against the fence with me. He scoffed and said, " My wife Ruth and my daughter Amelia, they separated us when we arrived. Allow me to introduce myself, my name is Francis." I shook his hand and said, "Nice to meet you; I'm James, James Walker." Hours went by as the day was slowly turning day into night, and I figured my time would go by pretty slowly and uninterruptedly as long as I didn't bring attention to myself. I more spoke with Francis a bit more; he seemed to be somebody I could trust and a man I wanted to help in any way I could. All of a sudden, I could hear two guards heading our way with shackles in one hand and the guard next to him holding his weapon. The same guard harassed me coming in; he swung the gate open and said, "The commander wants to talk; let's go, American." After I was shackled, I watched the guard close in, and he made it abundantly clear, " American you are going to die here, let me promise you that. It may not be now, but when it comes, I will take great pleasure in watching you suffer " I gave him a smirk and proudly stated, "And I'll take great pleasure in this," as I hucked up saliva into my mouth and sent it in his direction. Angry and caught off guard, he raised his rifle and smiled as he wiped my spit from his lower lip. "That doesn't mean you can't suffer a little bit first." He then proceeded to strike me in the stomach with his rifle, causing any leftover saliva to exit my mouth and the air in my lungs with it. He picked me up off the ground and pushed me forward.

Walking to the commander's office, I blindly tried to get the best look I could at the surrounding areas. Little holes in the bag positioned just right left me to notice that there was another gate entrance on the left side of the camp, and further down the road was a designated vehicle area where trucks would come and go. That would be my best chance of getting out of here, I thought as we approached the commander's building. "Wait here," said the guard as he knocked on the door. The building was three levels high, with about three guards on the outside and five to six more on each floor. I was brought up the stairs into the office and forcefully placed in the chair in front of the desk, and the bag over my head was removed. Out behind the desk was a highly decorated man looking out his window. Smoke rose from his cigar, and I awaited his next move. He said, "I am Major Klaus Zeigler, but you can call me Major Zeigler." Practically ignoring everything he said, I got more comfortable in my chair as if this were my office and said, "Listen, Kraut... wha..." I was interrupted by the guard's fist striking the left side of my face and falling over to my right. The guard picked me up, sat me back in the chair, and straightened me up. "You will show the Major respect, filth," said the guard with another fist ready if I didn't follow his commands.

Bringing myself back and looking at the Major again, I sarcastically said, "My bad Major, uh, Zeigler was it? Yeah, what's this about? Y'all, I just got here; I am not that important." Silent at first, he sat in his chair behind the desk and said, "You know, normally, when it comes to military prisoners, we would not bring you here. You would have been shot dead as you lay on the ground soaked in your blood. However, we feel you may have something of value to us." He turned around and put out his cigar in a silver ashtray. This office was nice with leather furniture and high- quality liquor and tobacco products. He sat in his chair and grunted a sigh of release, almost like he had had a long, tough day. Of course, this didn't seem likely from what I was looking at. "What the fuck do you mean, value?" I said, completely ignoring my advice to not attract attention and remain calm. Once again I was struck in the jaw, and the guard now filled with much more anger commanded me, "Watch your mouth; I swear I'll shoot you." My sarcasm only grew larger and I jokingly stated, " Hey asshole, you shoot me and the Major here won't get what he needs, and you hit me one more time and I reckon you'll get that ass beating you've been so well deserving of since I got here. you'll get that ass beating you've been so well deserving of since I got here." He glared at me with hate, nearly turning as red as a freshly bloomed rose.

Major Zeigler gave the guard a silent hand gesture and dismissed him from the room so that he could speak to me in private. The guard had exited the room, and the major stared at me for a second, then got up from his seat and paced the room until he was in front of the desk. He sat down and crossed his arms and looked at me and said, "Brave man, messing with the guard right then; man's killed more people than a battalion." My patience ran low, and the anger filtered in as I said, "You don't know me; you don't know anything about me." He slightly chuckled as he went to the back of his desk and pulled out the pouch of letters and papers in them. He opened the envelope where I kept my dog tags and raised them to meet my eye line. "On the contrary, Sergeant James Walker, we've read your letters and know you're a paratrooper with the... 509th Airborne? That's quite an achievement, no?" He dropped the letters and papers in the drawer and slammed it shut. Fiddling with the tags as we walked my way, he said, " I've heard plenty about your unit and how you've been Germany so much trouble. However, I will not let your malfeasance to our cause go without consequence. You know you put your whole unit in harms way, rather selfish if you ask me, no matter.". He walked behind me and slammed his palms on my shoulders. He leaned towards my right ear and me, whispering, "So tell me, James, where is your company headed now? See, I troops ordered you to ambush the oncoming troops, and your unit happens to be the next set of trash heading this way. Now, seeing that I can't possibly let them gain any knowledge on this camp, it would be much appreciated if you helped me out here." I gave a small laugh as I leaned in the other direction.

He could not be serious, and at the same time, his facial expressions gave him credibility. I kept my eyes forward and proudly said, "Well, Major, I reckon they are halfway to go fuck yourself-ville; your men may want to leave now so you're not late; oh, and your breath smells like cow shit; could you back up please? "He gave a gentle sigh and tapped my shoulders twice as he headed towards a knife displayed in front of his desk. He picked it up and informed me, "You know this was a gift from the Fuhrer himself; I didn't meet him, but his second highest ranking general delivered it; it was given to me as a sign of excellence and leadership running this camp. I don't know; I guess he was impressed or something.

Anyway, I digress; if you look closely, you can see rusted blood spots on the blade; it was used for those who proved themselves to be difficult." I wasn't exactly sure where he was going with this, although I knew no good would come out of it. He stepped to the right side of my chair and glanced down at me and told me, "I will get what I need where it is in red from your annoying, disrespectful, arrogant little body". He smirked, and with the emotion of absolute rage in his face, he brought down the knife with brute strength and pierced the top of my hand until four inches of the blade were seen from the other side of my hand. The excruciating pain caused me to scream, shed tears, and painfully yell, "Aah, you goddamn son of a bitch!" Holding onto the knife's grip, hesitating to pull it out, he placed his hand on me. " Oh yes Sergeant, I will get what I need from you, even if it means I have to cut you limb from limb and break all the bones in your body." He pulled out the knife and wiped the blood off on my clothing. Frantically exhaling air, I covered the open wound, and Major Zeigler snapped his finger, giving the signal that the guard was okay to take me back. He came in with the overhead bag at the ready. Zeigler turned to me one last time and said, "Sergeant, we will speak again, and next time I won't be so generous, I would think real hard about what you want to do. Real hard, get him out." The guard put the bag over my head and struck me with the stock end of his rifle into my lower abdomen and escorted me out and back to my area. Francis was resting on the fence again when he noticed the guard escorting my broken body back. The guard opened the gate and threw me in, and I fell to the ground. Francis kneeled beside me, "James? James? Are you okay? What happened to your hand? "I tore off parts of my pant leg and used it to wrap my hand and prevent further bleeding. I grunted as I sat up to speak with Francis. "I, uh, I met the Major. Yeah, real nice guy," I said sarcastically, tightening my makeshift bandage. "What did he say to you?" Said Francis in a worried tone. "I pushed myself back to the fence and leaned against the wire. "He said I was of value a." Francis looked away from me and looked towards the other prisoners in the yard and sat with a confused look on his face. "Let me look at it; I was a doctor back home," Francis said as he examined my palm. He turned my hand in both directions and said, "You'll live." Days had passed, and my concern for Bates being successful in finding me had risen, and I began to think that I may never see Leah again. Praying to the Lord and staying calm is going to save me. I just need to focus, I thought as I paced the yard. The next morning, I was panicked as I woke up because of the gong their guards used to wake the prisoners. It was also used as a way to stay on schedule. With each gong crash, it was meant to be our sign to move on to the next scheduled activity. For example, the second crash would mean it's time for chow, using the restroom, or making coffee if you were lucky enough to find some. This morning, as it went off, I thought a B-52 had dropped a bomb on the camp. I started my way up and looked through the fence, hoping it was

Bates and then men coming to my rescue. Sadly, I was disappointed; it was only the next batch of prisoners coming into the camp. I used the dirty, poorly manufactured latrines they had to offer and relieved myself before heading back to get my food. As I was in line, I could hear guards saying, "Knock it off you shit," or others saying, "I will shoot you in the head; stop squirming." I just figured it was some asshole trying to prove how tough he was and showing that the Nazis weren't going to have this easy. I turned back to the table, ignoring the commotion going on at the gate until my heart sank, my legs trembled, and my eyes opened as wide as they could before popping out as I heard the sentence that would change my entire mindset: "I'M THOMAS WALKER, YOU SUNZABITCHES, I

WILL KILL ALL OF YOU KRAUT FUCKS!" Completely unhinged because I had thought I lost my brother in the war. I thought I would be happy to see him, but instead I was annoyed. I threw my hand in my face and whispered to myself, "Of course, dumb motherfucker." Francis came over chuckling to himself and said, "Damn fool, like that'll help you. Who does he think he is anyway? Jesus? " I lowered my hand from my face and put it on the fence as I reluctantly told Francis," Nope, that's my brother, the biggest damn moron in the world." Thomas would pull one too many punches, and the guards had enough. The guard with his left hand slammed his rifle across his face, and before he knew it, he was knocked out and fell to the ground. They dragged him by his arms, brought him over to our living area, opened the gate, and threw him in. I didn't want the guards to know that I knew him in any way, so I left him to sleep on the ground, and once he woke, I would make a move to tell him I was here. Thomas would be out for a couple hours until the next scheduled gong would wake him up for dinner. He got up slowly and rubbed his hand across his jaw in an attempt to ease the pressure and pain the rifle had left. A couple of Jewish prisoners would help and inform him that it was time for dinner and we had to get in line. Once he was over there, I told Francis, "Okay, I'm going to try and talk to him; y'all keep watch for me." I slowly went around the other prisoners and gently moved them out of my way by tapping their left arms. Once I was next to him, I had to fight the urge to hug him and shed some tears of joy, finally meeting someone I had known before my imprisonment. I kept my eyes forward and back turned to the guard as I whispered, "Tom, keep looking to your left; don't look back." Angry and nervous, he said abruptly, "Fuck off, I don't want to hear anything y'all have to say to me." I turned back to see if any guards were looking and tried to talk to him again, letting him know it was me, "No, no Thomas, it's me, it's James." He dropped his tray on the table in front of him, which caught the guards' attention, but they ignored it and turned away once more, instantly looking in my direction and with a pale face.

Halting the line, he nervously said, "Jay? Wha... how? Uh, Jay, how did you get here? Why are you here? My God, I thought I lost you." I grabbed him by the arm and reassured him that I was okay, just a little beaten up. We grabbed our food and went to sit at one of the tables in the yard. I told him that I had been taken in Saint Lo, and the bombs blew up the bridge before I could cross. " I tried to escape but one of them German officer fucks captured me before I could and brought me to this shithole. I reckon you and I will be given the most attention; it seems the Major knows us." He put his fork down and leaned in closer so our conversations weren't vulnerable to those we couldn't trust. "Why? It seems to me like Nazis could care less about you and me," Thomas said, looking to his left and his right. I nodded my head and couldn't give him an answer. Thomas leaned back and resumed eating, and after a few bites, he asked with his mouth full, "Okay Jay, so what's our play here? You reckon we die here; I'm not too attracted to that idea." I looked around to see if any guards were listening in and said, "I'm not sure; I saw that they have vehicles near the back entrance; trucks come in and out every day; the only problem is, it's heavily guarded; we'd make it ten feet before they put holes in our back.

I have a man from my unit coming up with a plan to find me; I suppose all we can do is wait until they come."

Thomas was not pleased with this idea and gave me a disappointed look; however, he agreed our options were low. After dinner, we rested in the yard until we heard yelling from the other fence. "So you want to steal food? Okay, commanders have a plan for thieves like you; come here! "The guard had grabbed a young lady, who couldn't have been older than twenty-five. She had stolen a dinner roll from the table to save for later. I couldn't say I blame her; these meals they were feeding us weren't exactly gourmet, and the portions were even worse. The guard began pulling her by the hair and took her to an old, rusted building around the corner. The fumes were pungent and had another smell of rotting corpses. We all watched, knowing we couldn't do anything to help her—at least nothing that wouldn't put us in the same position. The guard threw her in the room, went in behind her, and locked the door, with the major going in after.

After only a few seconds, we could hear the traumatizing screams coming from the women. She was begging and pleading for them to stop, and soon after, her screams were silenced. The guard and the major walked out, but the young lady did not. Major Zeigler walked to the center of the fences and made an announcement: "Rules will be followed or you will be subject to the same punishment, that there is a gas chamber, you will be locked in, and you will be in there until you die, like the garbage you all are." He looked around with anger and then left back to his office. Prisoners were not worried; they have all been here for some time and have seen many meet the same fate, almost as if they expected to be next. These examples were nothing new, as the Major would come out daily to execute another innocent life. He would begin by walking by prisoners, scoping out a man, woman, or child that would bring great significance to his message. Thomas and I witnessed as he halted and spotted an innocent mother and her son.

Without words, he pointed and commanded the guard to bring

them to the wall between blocks ten and eleven. Here was an alley where a wall cut off the back, and they would line up the innocents, strip them of their clothes, and put them on their knees. "Everyone watch, as this will be your fate—maybe not today or tomorrow, but mark my words, your time will come." Major Zeigler turned back towards the mother and child and nodded to the guards. They raised their weapons and fired upon the boy, the defenseless boy, and his mother. The screams from the prisoners brought joy to the Major; it seemed that he felt that this was his purpose in life—this is what the good Lord created him for. "You coward, they were innocent; you're a bastard and a man of weakness, praying on the weak. You'll pay for this," said Thomas, as tears fell from his eyes. I looked at Thomas, pulled him off the fence, and said, "Tom, what are you doing? Shut up." The Major stared towards the dead bodies for a minute and then looked our way. He approached the gate and said, "Bring me the Americans; I think another conversation is in order." The guards rushed in, and Thomas put up a fight. He struck the first guard in the face and rushed towards the other but was stopped by the guard who had tackled him to the ground. More guards came in and brought us to Major Zeigler's office. We were seated in the same chairs I was brought to days earlier. Zeigler lit a cigarette and said, "That was foolish of you American; normally I choose one to two Jews a day to execute, but because you can't keep your mouth shut, I'm going to make it five per day." Thomas had difficulty controlling his anger as he was a man of justice and sought to defeat those who pursued evil, especially those who had done nothing wrong. He looked at the Major and said, "Fuck you, kraut, I'm going to enjoy taking your life, not with a gun, not with a knife; nah, I reckon my own bear hands will satisfy me, choking the life from your body and pissing on the corpse. How does that sound, Major?" The guard had walked back in with urgency. "Commander, we found his belongings; you'll want to see this." The Major looked away from Thomas to see what the German guard had brought. It was Thomas's notebook and his dog tags.

"Sergeant Thomas Walker... Walker, well, isn't this sweet, brothers?" I whispered under my breath, "Fuck." The Major grinned as he walked back behind his desk to take a seat. "I told you, Walker, I would get what I need, even if it means I have to break every bone in your body; however, I think I found another way." Zeigler opened the drawer to his left, pulled out a hammer, closed the drawer, and walked behind Thomas and me until he was on his left side. "I'm going to ask you again, Sergeant. Where is your unit, and what's their next target? I suggest you answer me this time, for your brother's sake." The pressure was settling in as I knew I had to choose between my unit's safety and my brothers. I couldn't imagine the thought of losing my brother or possibly losing Bates. "I want an answer now, Sergeant," said Zeigler as he held the hammer over Thomas's head, ready to swing. I shook my head in disappointment as I knew I had to choose. I turned to Zeigler slowly. "Alright, Major, alright, their..."NO! "Said Thomas as he didn't think his life was more important than the dozens of American lives planning a rescue. Major brought the hammer down as hard as he could onto Thomas's left hand. Thomas screamed in agonizing pain, with tears coming from his eyes. "You think this is a game? I do not need your brother; his life means nothing to me, and I would find pleasure in killing him." said Ziegler as his emotions had overpowered him and he thought killing Thomas would probably give me more motivation to answer. I stood up, ready to fight, but the guard brought me down as he struck me in the stomach. "Sit the fuck down," yelled the guard. "You son of a bitch, coward, he has nothing to do with this." Zeigler threw Thomas back and onto the floor, causing him to fall out of his chair, clearly frustrated by the fact that we were stubborn and arrogant. "Perhaps you need a little bit more motivation, Sergeant." Zeigler turned to the guard holding Thomas and said, "Get the bag; put it over his head; I think he's thirsty." Holding down Thomas's arms, the bag was placed over his head, and the Major grabbed a water pitcher. "Now you're going to tell me, and I would suggest being truthful; pour it." The water began to fall, and the breath from Tom was struggling to remain in his body. I could hear him choking, and the sounds of him being tortured caused me to break. "Alright, that's enough; no more, no fucking more, I'll tell you, just stop alright." Zeigler stopped pouring the water, and the guard sat Thomas back up in his chair and took the bag off his head. The major put the pitcher back on his desk and was ready to listen as I told him, "Last time I saw them, they were in Carentan; I haven't seen them since; they could be across the damn country by now."

Appearing disappointed, he walked behind his desk, opened the drawer, and pulled out a Luger pistol. Zeigler walked back to Thomas and raised the pistol to his head. "You must like lying to me; you must not care about your brother. Shame." Cocking the pistol, he applied pressure to the trigger. "Wait, wait, I swear I'm telling you the truth; I haven't seen them since then; I was blown up by you Kraut fucks; they captured me and brought me here; that's all I know, I swear." The major lowered his pistol and threw it on his desk. "Alright, Sergeant, but if you're lying to me, I will kill your pathetic brother and I will fucking roast you in the chamber. Am I clear? "Pissed off and ready to kill this bastard, I responded, "Crystal." He went back and sat in his chair and lit up a cigar, looked at the guard, and said, "Get a team ready, and head to Carentan; pay the Sergeants unit a visit; no loose ends; now get these fuckers out of my office." We were escorted out of the office, and I was terrified that I may have put the lives of many in danger.

CHAPTER 14

STOP THE GERMANS OR DIE TRYING

Tensions were high back at Carentan as the word got out that I was captured. Bates was with Captain Jacoway and Captain McDonald, and the three were trying to configure a plan to rescue me. "Sir, why don't we just get air support and blow those Kraut fucks to hell?" Captain Jacoway was at the strategy table and placed his hands flat. "We can't do that, Bates; we don't know what their forces are like, and if we have civilians at the camp, we can't risk the casualties." Understanding but frustrated with the captain's answer, Bates loudly exclaimed, "Well, dammit, sir, we can't just sit here and do nothing; y'all need to get off your ass and take charge for fucks sake." Captain Jacoway immediately put Bates back in his place and tried to calm Bates down. "Bates as you were, you watch your fucking mouth when you speak to me, do you understand me?" Bates turned and remained silent. "Bates, do you understand me?" Bates looked up at the captain and made eye contact.

"Yes, sir, I'm sorry. I'm just close with Walker, and I fear we may be too late," he said, knowing he was in the wrong. A few minutes passed, and everyone spoke their minds on a solution to rescue me. With no reasonable solution, the captain said, "Alright, y'all, we aren't going to have a good plan with our minds the way they are; turn in, get some rest, and we'll plan and move out in the morning. Goodnight, gentlemen." Emotions were high, and the stress was so overwhelming that Bates couldn't rest for a second. Shadowed by guilt, he felt as though my capture was his fault. Pacing the ground and constantly reminding himself that he could've pulled me out and brought me to safety, of course I knew that wasn't true as I was well within the blast radius of the bombs, but he just couldn't let that go.

The next morning, the captain's called everyone to the town center to devise a strategic plan to help me. "Morning all, ok, first things first, we need to know where the Kraut bastards took him," said Captain Jacoway. Bates had no clue of where I had been taken, he glanced around and noticed the German POW's and told Captain McDonald, "Sir, we can ask the Germans, maybe they know which camp he was brought to, I mean they have to know something right? I say it's worth a shot. Let's hope to God they know something otherwise the Walkers might be good as dead, and the other innocents to. Captain McDonald decided it was time to interrogate one of the German POWs to see if they knew. Captain McDonald ordered two of the soldiers to grab two of the Germans. They brought them to where we were meeting, pulled up some chairs from the debris, and sat them down. "Translator? Translator, come here," said Captain Jacoway, as we needed him to ask them the question. The translator hustled over and was ready to translate. "Ok, ask these fuckers where the camp is." The translator turned to them and asked them the question in German.

Loyal to their country, they laughed and ignored the captain's question. Captain Jacoway and McDonald were not so amused, and with low patience, he pulled out his pistol and pointed it at the German on the left and said, "Ok, let's try this again; ask them the question again and tell them if they ignore me again, I'll kill this motherfucker; go on." The translator asked them again, and the German on the right said, "Fuck you," in German. Jacoway chuckled as he meant it when he said he'd shoot, and he raised the pistol to his head and pulled the trigger. The German to the right stopped laughing immediately; now his eyes were filled with fear. Jacoway shoved the dead German's chair over with force and knocked him to the ground. Afterward, he grabbed the German by his chest and said, "Alright, last chance; you're not the only shit stain I can ask, so don't fuck with me." With more anger, he told the translator to ask him again. He did, and without hesitation, he gave up the information. The translator turned to the captain's and said, "He says that it's a camp called Auschwitz; he said that he's probably dead already; they don't take too kindly to Americans." Captain Jacoway nodded and then shot the German in the head, as he was no longer necessary to him. Captain Jacoway ordered, "Alright men, get your gear and be ready to move; pack light; only what you need for a fight; we're getting our men back." The men went to prepare, and Jacoway pulled Bates over to speak to him privately. "Alright Baters, listen, this isn't going to be easy. I know you and Walker were close, and I get it; he's a good man. Understand that we are walking into hell's fire, we may not all make it out." Bates nodded as he understood the risk and was ready to sacrifice everything to save me. He turned to Captain Jacoway and said, "I know, sir, thank you, and I'm ready to go." About an hour later, they moved out and headed my way.

Meanwhile, I sat next to Thomas and Francis at the camp as we recovered from our last visit with the Major, still worried that nobody was coming for us. We heard commotion as the Major conjured up an army to head towards the platoons' direction. He made a quick stop my way, looked at me, and said, "Make no mistake, Sergeant, we will overpower your pathetic American friends; you will die here, you and your brother. I think the chamber would be a nice touch; that may be too easy for garbage like you." Zeigler chuckled and returned to his office as he gave the order to march out. The Germans marched out singing a war song; they were ready to kill, and they had no plans on capturing anybody. While the camp was near empty due to most of the guards joining the Germans going to fight, they left a couple to maintain the Major's standards. Every so often, they would select somebody at random to carry out the killings with the chamber. Watching innocents get murdered took a toll on my mental state. I rested on the fence and laid my head back. Thomas looked at me and asked, "Jay, are you good? What's up, man?" I opened my eyes, looked up, and said, "I'm just worried about Leah; she must be worried sick about me, and I reckon she thinks I'm dead if I haven't sent a letter in this long. At any point, we could be put in that chamber, and I'll never get to tell her how I truly feel, you know? Ain't there no way we're making it out here?" Thomas also rested his head against the fence and sarcastically said, "Well, aren't you a sour taste in the mouth?" Major Zeigler left his office and walked to the prisoners across from us. He pointed at Francis's daughter and ordered the guard to bring her to his office.

The mother tried to fight off the guard, but he struck her across the face and took the daughter by force. He escorted her to the office. Francis began yelling, "No, Amilia! You bastards, let her go, Amilia!" Francis tried to break through the fence but was halted as the guard aimed his rifle at him. The mother had snuck through the fence and began running towards the daughter. She was caught by one of the guards and brought to her knees. Zeigler stepped out and approached the mother. The Major stared at her for a second as she begged him, "Please, not my daughter, please, I'll do anything; take me; sir, please take me; I'll do whatever you want; just let her go, please."

Showing no sympathy for the weeping mother, he pulled out his luger and aimed it at her head. The daughter cried, "Mama, no, please." She smiled at her daughter for the last time and said, "Its ok, baby, its ok." Zeigler put five pounds of pressure on the trigger and left a lifetime of torment in Amelia's life. She hit the floor, with no breath and a bullet wound through her forehead. She was killed right then and there. Francis dropped to his knees, speechless and frozen. He began screaming and crying, as there was nothing he could do. The mother's body was dragged to the dead body pile in the camp, and the daughter was forced into the office. She was hesitant at first as she tried to escape his grip, but Zeigler raised his hand and struck her, and she fell. Francis yelled at the Major, "You sun of a bitch, I'm going to kill you, and I will kill you, bastard." Zeigler looked at the guard and nodded, giving him the silent command to do as he saw fit. The guard opened the gate door and grabbed Francis from the ground. Although he fought well, he was no match for the three guards that had grabbed him. They took him to the chamber, and we never saw him again. Thomas and I just watched as they grabbed him, for we couldn't do anything that wouldn't land us in the same situation as him. Thomas whispered to me, "Don't worry, Jay, we're going to get that Kraut fuck and we are going to fucking kill him, you hear me? I reckon he'll get what he deserves." Hours passed, and we noticed a guard escorting the little girl out of the office. Zeigler stared at me as he pulled up his zipper. Amelia was left traumatized and what we could only imagine to have been the worst experience in her life, she stumbled down the stairs and was placed back into the holding area. I turned my head in anguish and thought, Come on, Bates, you gotta help us.

Captain Jacoway and McDonald continued to lead their men for a couple of days until they decided to rest before making the final push to the camp. Jacoway sent a man onward to pull security and run back if he noticed any Germans heading our way. The night was quiet, and Bates and Ortiz rested on a tree trunk, ate their meat can tins, and drank some water. The captain came over, rested against them, and conversed. "Tomorrow's the big day, alright? We make our final push to the camp. Bates, if Walker didn't make it, then we're going to avenge him and give them hell. Justice will come, and that's a promise, ok?" Bates frowned downward and nodded. Captain Jacoway got up and turned in for the night. Bates said a little prayer before he fell asleep: "Our Father, Who art in heaven, Hallowed be Thy Name. Thy kingdom come, Thy will be done, on Earth as it is in heaven. Give Walker the strength to overcome; let us be healthy and allow us to perform your work; give me the strength to bring him back; let him see his girl again; help us to victory; and if we are to fall, save us a spot in heaven, for we are your servants; for in your name I pray, Amen." Bates rested his eyes and went to sleep.

Morning rose, and the runner came sprinting back, yelling, "Their coming, the Germans are coming!" Captain Jacoway stopped him and said, "Woah, Private, slow down; who's coming?" Catching his breath, he said, "The Germans are singing and marching here now; they look ready for a fight, sir." Jacoway looked in the distance, as he could hear the Germans approaching. "McDonald, do you hear that?" Captain McDonald looked in that direction and said, "Shit, they're coming; we need to get ready now." Captain Jacoway looked nervous and shouted, "Alright men, get up, let's go; it's time; ready your weapons and take defensive positions; fire on my command." Bates and Ortiz posted behind the tree for cover and aimed their weapon downrange. He removed his weapon and whispered to himself, "For Walker." The Germans came over the hill, and Jacoway ordered a fire. He released the first shot, and the soldiers followed right behind him. Before they knew it, the firefight had begun. This would be the final stand before reaching the camp. The Germans immediately began taking cover in a depression below the hill. They laid down, set their MG42's up, and fired at us.

Within seconds, their weapons took out nearly ten men who were behind poor coverage. Soldiers began running back to take cover behind some trees but were cut down by the rounds heading our way. Captain Jacoway yelled, "Behind the trees, behind the trees, Bates, you grab Ortiz and lay down heavy fire on those MG42's right now, go!" They sprinted towards the nearest divot in the ground to take cover and focused their fire on the 42's. Ortiz popped his head up and took three shots their way. He had hit the feeder of the gun, and the German firing took cover. With the gun down, the platoon advanced for better coverage. "Grenades, grenades, on my count," said Bates as he had one prepped. "Ready, one, two, three, now." They tossed the grenades over and hit the German on the gun. More troops pushed their way towards the soldiers, throwing their grenades. As one soldier was heading to Bates, the grenade went off and took both his legs; he died instantly. The battle raged on, and Captain Jacoway and Captain McDonald knew they would be sitting ducks waiting to be slaughtered if they didn't move. Deciding it was time to push forward, he told nearby soldiers to throw smoke grenades in front of their position to provide concealment so they could move up.

Desperate to move, two soldiers found themselves to be unfortunate casualties as they ran as soon as the smoke popped. Another MG raged at them, cutting them down in their tracks. Jacoway said, "Fuck, wait for my command god dammit, ok wait...wait...NOW, move, move, move," They shot through the smoke, hitting the Germans with blind bullets. Both sides were losing men equally, but they knew they had to keep pushing to overwhelm the Germans. They were halted in their advancement due to the MG fire. Bates looked at the captain's and said, "Captain Jacoway, Captain McDonald, we are good as dead if we can't get past those guns. I'm going to flank left. I'll take Ortiz and a couple of men. You two distract them down the middle." Captain Jacoway gave the okay, and Bates threw another smoke to provide him cover. Bates and Ortiz pushed left, and the men behind them joined. They ran to the divot in the ground where the Germans were firing from a.

The Germans were falling one by one, and the Germans were being pushed back. The soldiers advanced once more, and the German troops were dying off by the numbers. Accepting their defeat, the Germans began surrendering. Captain Jacoway was ruthless, as prisoners were of no exception; they killed them as they raised their hands, saying,

"Spare no one; they don't deserve mercy; keep going." Captain McDonald told Jacoway, "You know we can't fire on those who turn their backs, right?" Jacoway looked at him, said, "Yeah, I don't care," and watched the Germans fall. Loyal Germans who stayed gave a good effort as they killed a dozen more U.S. soldiers, but their efforts would fall short as Bates kept firing his Thompson machine gun and killing them. They began retreating but would be killed by our forces before they got too far. The Germans had been defeated, and Jacoway gathered the remaining men and said, "Men, you fought hard, and this battle is won. It's not over yet. We move to the camp, get our men back, and save civilians. Now let's get this over with."

CHAPTER 15

NO LOOSE ENDS

Unaware that his troops had been defeated, Zeigler had the guards' escort Thomas and me back to his office. Once again staring at his window, he said, "Do you know why we use the chamber? Hm? We could easily just line up all of you and shoot everyone, but that doesn't send a message." Thomas and I stared at each other, wondering why he was telling us this. What could we gain from it? "The chamber acts as a guideline if you will, keeps all the prisoners in line, and lets them know that rules are meant to be followed and disobedience of any kind has consequences." Thomas looked at Zeigler as he turned around and asked, "By murdering innocents? That's how you send a message? It seems like a coward's way of killing those who don't have the strength to defend themselves. And what about that girl, you sumbitch? Raping her, was that a message to her?" Zeigler sat in his chair, put his hat on the table, and eased back. "Asserting dominance is how I control these people and tell them I am their new leader, their God, as I like to think about it. And let me make it clear: I will continue to show these pathetic, lifeless, garbage rats that they will always be beneath my boot; their lives are now forfeit." Disgusted by his answer, I waited for why he brought us here. "Fear not the Fuhrer anymore; he has taken his life—a cowardly act, wouldn't you agree, Thomas? Leaving us with nobody to lead, I now see it as my responsibility to show these prisoners that there is still a man who delivers, true to his word." I sat up in my chair and had to mention how crazy he was to fight for a cause that was not feasible anymore.

Confused, I asked him, "So what do you plan on being the next Hitler? I reckon it doesn't make much sense considering you're a damn near-fallen country." Zeigler grabbed a cigar from the right side drawer, lit a match, and puffed before responding to me. "You simply don't understand classic American arrogance. I wish not to be the next Fuhrer; that's silly. No, I wish to be an example of true power. I will not rest until every prisoner here sees what true power is. I will execute every fucking person here and piss on the ashes," Zeigler said as he began to stand from his chair. "I will show you what being a leader means." Gunfire interrupted our conversation, and the two guards outside the door rushed in to give the news. "Major, we are under attack; the Americans must have defeated our troops." Zeigler grabbed the lamp on his desk, threw it against the wall, and said, "What? Impossible, watch these two; if any of them move, you shoot them; get out of the way." The door was closed, and the guard aimed his rifle at our backs with the finger on the trigger.

Thomas and I knew we couldn't make any sudden movements as he was shaking, and any movement of any kind would cause a bullet to be put in our backs. When Major Zeigler stepped outside, he saw that Bates and the two captains were leading troops in an ambush against the camp. Zeigler immediately ordered his men to return fire and take their positions. MG's were loaded, and the guards used the crates and corners of buildings to take cover. The Germans were using the second and third floors of the building to fire from higher ground. I could hear the voice, which sounded like an angel at the time: "Don't worry, Walker, we're coming; just hang on." Bates said as he rushed through the front gate and kneeled behind a barricaded crate. I turned my head backwards towards the door, as I couldn't believe they found us. Zeigler retreated to the tower that was towards the back of the camp. A German was taking cover as bullets hit, and Zeigler grabbed his Kar98 sniper rifle and began firing at the soldiers. Jacoway looked at a nearby soldier and told him to clear the building next to them. The soldier stood up and said, "Yes, sir, moving." Before he made two steps, Zeigler pulled the trigger, and the soldier was struck in the neck and fell to the ground. Zeigler would take three more men out seconds later. "Sir, I'm clearing this building. I'll look for Walker. Cover me." Captain Jacoway made the announcement, "COVER FIRE, Go!" They stood up and shot at the tower. Zeigler took cover, and Bates moved in as bullets pierced the building's wall. He busted through the door so hard that he fell to the ground.

He was met by three Germans firing out the window. Bate's stood up, shot the first German in the chest, and grabbed him for cover against the other Germans bullets. He grabbed the luger from his holster and fired at the other two, killing them. Bates could hear Germans firing from the second floor and proceeded up the stairs. He reloaded his Thompson and cleared the second floor. He shot the first German and sprinted towards the second before he could get a shot. Bates used his weapon to shove him through the glass and out the window. As the German fell to the floor, soldiers from down below shot the body. Bates looked out the window and saw the civilians in the fenced areas and shouted to the captain, "Sir, civilians just past these buildings; we have to get them out of here." Jacoway acknowledged and told Captain McDonald he needed to lead some men to the right and focus on getting civilians out of here.

Captain McDonald turned to the men behind him and said, "Alright, all of you with me; our goal is to free the prisoners and get them to safety; let's move." They rushed to the right, met with returning fire and pushed to the holding areas. Ortiz told Bates that he was pushing to the second building. Ortiz yelled to Bates, firing from the window,

"BATES! I'm moving to the next building; meet me over here." Bates nodded and provided covering fire. Ortiz fired shots downrange and killed two Germans hiding behind the crates. He entered the door, where he noticed four Germans firing from the window. He fired on them and killed three before he ran out of ammo. Not having enough time to reload, he threw his rifle at the last remaining German in the building and had to fight him in hand-to-hand combat. Ortiz struck him in the face, causing him to fall to the floor. The German struck him in the nose, forcing him off and causing his nose to bleed. The German then pulled out a knife and paced around him. Shouting, he rushed towards Ortiz in an attempt to stab him. Ortiz grabbed his arm, and the German tripped him, and they fell to the ground. The German was on top of him and trying to force the knife into his chest. "Hold on, Ortiz, I'm coming," said Bates as he exited the building, firing to the house. Shots stopped him from the tower as Zeigler was sending rounds his way. "Ortiz, Ortiz, hang on." Exhausted, Ortiz was relieving pressure from restricting the knife. The German was putting in all of his effort and began punching the knife to help send the knife forward. He was successful, as Ortiz couldn't hold him off anymore and was pierced in the chest. The German kept pushing, and Ortiz was losing breath by the second. Zeigler ceased firing as he had to reload. Bates ran through the door, where he saw the German on top of Ortiz, before the German could grab his weapon. Bates fired shots at the German, hitting him in the chest and head. Bates rushed over to Ortiz and said," Ortiz, hey, Ortiz, stay with me, man. Come on, Ortiz, you're fine, you're ok." Reaching for his first aid, he grabbed bandages and applied pressure to the leaking wound. Ortiz was losing too much blood and grabbed Bates by the arm and said, "Get Walker; get him out of here. I'm sorry, Bates. I'm sorry." Ortiz took his last breath and died in his arms. Bates began to shed tears as he knew Ortiz was gone. He grabbed his dog tags and then posted up on the front door, he glared at the dead body of Ortiz once more and felt nothing but guilt rush through his body. "I'm sorry Ortiz, I wasn't fast enough, these fuckers are going to pay for what they did to you, I'll be back for you brother," Bates said as he made his Thompson ready. Bates looked at Ortiz and yelled for Captain Jacoway, "Ortiz is gone; we need to move now." Captain Jacoway had his men advance, and they pushed on to the prisoners.

Meanwhile, Captain McDonald was fighting off the guards with the prisoners. McDonald stood up from his cover and shot two Germans crossing the road. He dove and reloaded his weapon. Zeigler noticed troops coming to the prisoners and saw McDonald's helmet peeking from the crate. He took a slow, inhaled breath, and as Captain McDonald stood up to fire at the guards, Zeigler fired, hitting him in the left shoulder. Captain McDonald fell and yelled; "SNIPER! Stay low." His men were now stuck behind cover as the guards moved to better cover.

Thomas and I were in the office when we heard the chaos outside the door. Thomas looked at me and whispered, "Jay, we gotta help them, and we need to take the fuckers out." The Germans heard Thomas whispering and told us to be quiet as they stiffened their rifles. I looked at Thomas and gave him a nod; he knew I was hinting to wait for my move. I began yelling at the guards, "You cowards should be out there getting slaughtered with the rest of them, but no, y'all are hiding in here." The German to the left approached me and struck me in the face. With him being close, I grabbed his rifle and began fighting him. With the other guard aiming my way, Thomas stood up, grabbed the rifle, and pushed him into the wall. I delivered strikes to his face, grabbed the rifle, and began hitting him with the stock until he died. Thomas, fighting the guard, yelled my way, "Jay, shoot this fucker!" He struck him in the stomach and pushed him away, and that's when I raised the rifle and shot the guard. We took their weapons and ammo and made our way outside. Thomas and I began firing at the guards positioned towards our troops and shot them in the back. Captain McDonald yelled to the man carrying a rocket and shouted, "Hey! Hit the tower; get his ass off of there!" The private aimed that way and sent the rocket hitting the wall in front of Zeigler. He was forced off the tower and made his way to the gas chamber to take cover. Captain McDonald met up with the rest of the platoon. Captain Jacoway asked, "Mac, are you good?"

Captain McDonald said, "I'm good; I'm good; just a shoulder wound; I'm in this fight; come on, let's get these prisoners out of here." They were all lying on the ground, trying to avoid the bullets, and Captain McDonald and his men opened the gates, and the prisoners began running out the front gate to freedom. Many prisoners, in the heat of rage, picked up weapons to help us fight the guards. With the Germans retreating, they were met with fire from troops and prisoners. Only a few Germans stayed, as they fired back and killed some of the civilians. We cleared the front of the office, and I heard a voice say, "Walker, walker, It's me. Its Bates." I turned the corner where I saw a rifle aimed at me; it was lowered, and Bates was right there. With a look of surprise on his face, he ran towards me and wrapped me in a hug, saying, "Holy shit, I thought you were dead. I told your ass I'd come for you." Relieved to see him, I responded, "Bates, it's so good to see you man I can't believe you guys found us, and this is Thomas, my brother." Bates shook his hand and said, "Good to meet you." Bates said, "Look, we need to get to Captain Jacoway; let's go." I told him, "Bates, I have to get the Major; he dies today, and he went into the gas chamber. Thomas, are you ready?" Thomas said yes and cocked his weapon, and we all ran to the chamber.

With the Germans returning heavy fire and the non-fighting prisoners escaping the camp, Captain Jacoway shouted to Captain McDonald, "Hey! I'm calling for air support! we're going to blow this camp to hell. Everyone, make your way out of the camp. Go, go, go!" Captain McDonald agreed, and he and a few soldiers readied their smoke grenades to conceal themselves from fire again. Once clear, Captain Jacoway called over the radio, "Air one, Air one, this is Ground two, requesting immediate air support on my location. Again, I am requesting immediate air support on my location." Captain Jacoway gave the coordinates, and they radioed back, "Ground two, this is Air One; good copy; get clear; drop in four minutes."

They retreated out of the camp and took cover, ready for the bombs to drop. Bates, Thomas, and I approached the entry to the chamber and breached through the door. Once inside, we could see Zeigler inside, holding his luger to Francis's head. "That's far enough, Sergeant; don't come any closer or I will shoot him. Did you think I wouldn't have a backup plan? Now lower your weapons or I will kill him." Stunned by the fact that we thought that Francis had been killed, I was overwhelmed, and we kept our rifles pointed at Zeigler. Frustrated and out of patience, Zeigler fired a round past his ear, shouting, "I SAID DROP YOUR WEAPONS, NOW!? I told Bates and Thomas to lower their weapons. Thomas said, "Jay, are you serious? Let's shoot the bastard." I dropped my gun on the floor and told them to once again lower their weapons, and they did. With our hands in the air, I told Zeigler, "Ok, let him go; this is between you and me; just let him go." Zeigler stared at me and then aimed his luger my way and shot me in the right shoulder. Thomas yelled my way, "JAY, JAY, you ok, you ok? " Thomas tried to approach me, but Zeigler aimed the pistol his way and said, " I wouldn't do that; stay back.

You know, Sergeant, you are a real pain in my ass; you're getting in the way of my mission, and it's truly aggravating." Two guards came in and aimed their rifles at Thomas and Bates to secure them. Zeigler stated," So here's what's going to happen, you and your American trash friends here are going to lay on the ground, and I'm going to take a vehicle out of here. You try anything, and your friend here dies." Zeigler fell silent, as he could hear roaring engines in the sky heading this way. Air support radioed to Captain Jacoway, "Ground two, Ground two, this is Air one, dropping in three, two, one." The bombs fell and began

hitting the ground and buildings. A bomb hit the chamber and caused the building to begin collapsing. The impact caused Zeigler and the rest of us to fall, knocking the pistol out of his hands and allowing Francis to escape his grasp. I sprinted towards Zeigler and engaged in hand-to-hand combat as Bates and Thomas took care of the guards. The ceiling began to fall, and once Bates took out his guard, a piece fell and landed on his leg, breaking his leg. Bates was trapped underneath the rubble. The bombs killed the Germans outside, and the rubble blocked the U.S. troops out. Thomas ran to Bates and tried to lift the rubble off of him. Meanwhile, I began striking Zeigler in the face. I forced him to the ground and continued beating him. Zeigler grabbed a nearby rock and smashed it across my head, knocking me off of him.

He climbed on top and started striking me repeatedly. I grabbed his arm and locked his arm before his next attempt to strike. I pushed him off and stood back up. A second bomb landed near the chamber, and the center of the chamber's ceiling fell, blocking me off from Thomas and Bates. Francis took cover, and it was just Zeigler and I. "For years I successfully managed this camp; for years I operated this camp with no problems; then here you come, fucking it all up; I'm going to kill you, and then I will massacre every Jewish filth in the country." He pulled out the knife from his office holstered to his belt. I looked around and noticed a steel pipe on the ground. I picked it up and said in anger, "Fuck you, these people are innocent; you're done hurting people, Zeigler. I'm going to make sure of that." He yelled as we ran towards each other, and I swung the pipe, hitting him in the left arm. He dodged the next swing and swung his knife, grazing my arm and tearing the sleeve. I grabbed him and threw him to the ground. I attempted to swing again, but he caught the pipe and drove the knife into my leg. I fell to the ground, and he climbed on top. He struck me in the face; punch after punch, I was surely dead. He punched the knife deeper into my leg and continued to strike me. With each strike, he said, "Fuck you," the next strike said, "Fuck these Jews," and the next strike he said, "And fuck you American assholes." Zeigler pulled the knife out of my leg, and blood began rushing down my leg. With great force, he brought the knife down towards my chest. The building was still collapsing, and I had to move quickly. I blocked it with my arms and started to push back. The knife inched closer to my chest, and Zeigler said, "Goodbye, Sergeant Walker; your time is up." With all my strength going into resisting the knife, I saw a cement stone next to me. I looked back at Zeigler and said, "No Major, your time is up, you Kraut bastard." I grabbed the stone with my right hand and smashed it across his head, causing him to bleed aggressively. He fell over, and I climbed on top and began choking him with my bare hands.

Locking his arms with my legs, I squeezed harder and harder until I could see his soul leaving his body. Thomas yelled to me, "JAY, WE GOTTA GO!" Still trying to lift the stone off Bates, he yelled again, "Jay, come on, the building's going down; we've got to move now." I was blinded with rage as I kept pressing on Zeigler's neck. The ceiling above me was about to fall, and when I looked up, I could see a large piece of cement falling. I jumped off of Zeigler and dove away. The cement came crashing down, and before I knew it, it crushed Zeigler, killing him immediately. I stared at him for a second, as I knew his reign of tyranny was over and people were safe from him. Pieces continued to fall and knocked over the rocks barricading me from Thomas and Bates. Even though the bombs stopped, the building continued to fall. I rushed over to Thomas and began to help him lift the stone off of Bates. We lifted it off, and I grabbed Bates, rested his arm on my shoulder, and began to carry him out. Thomas was right behind me until he saw Francis in the corner. Thomas turned to me and said, "Jay, get him out. I have to get Francis. Go, hurry." Thomas picked up Francis and ran to the door. I got Bates out and rested him up against the wall. I turned back to Thomas to help him. The ceiling fell completely, and Thomas noticed and said to Francis, "Come on, Francis, go, go, go." The debris was falling too fast, and Thomas tripped over a large piece of debris and fell to the floor, throwing Francis out of the door. I yelled, "THOMAS, I'M COMING, HANG ON, THOMAS,

NO!" Thomas looked up as the debris fell and knew it was too late to save him. Accepting his fate, he turned to me one last time and said, "I'm sorry, Jay," and closed his eyes.

Before I could help him, the debris fell and crushed him in front of my eyes, blocking the entry. The world went silent, and I knelt before the chamber. I screamed, "THOMAS! THOMAS!" I started to move rocks out of the way and make my way to him. Once I got to him, I could see his body flat under the cement. I knew I was too late. I started to cry in agonizing pain. "No God, please no, fuck, Thomas, come on, come on, sun of a bitch." dried out of tears, I sat there with guilt and said, "I'm so sorry, Tom, I'm sorry." Captain Jacoway and Captain McDonald ran in with the troops and cleared the camp. They approached our position and saw what had happened. Captain McDonald grabbed Bates and Francis and escorted them out of the camp. Captain Jacoway walked up to me and put his hand on my shoulder. I grabbed a rock nearby and raised it, ready to strike. Jacoway said, "Woah, woah, Walker, It's me, Sergeant, It's me; it's Captain Jacoway; it's over, son; it's over; you're safe now; alright, come on, let's get out of here." I grabbed Thomas's tags, kissed my hand, and placed them on his head. Before I left, I said, "I love you, Tom. Cap, we need to get his body out of here." Captain Jacoway responded, "Ok, son, we'll get him; we'll get him; come on." Two soldiers removed the cement from his body and brought him out of the camp. Francis was reunited with his daughter, and the tears they shed with each other were heartwarming.

Hours passed, and as we waited for another unit to pick us up, we sat by a tree staring into nothing, feeling nothing but anguish and guilt. Francis came over to me and stood next to me. He turned towards the destroyed camp and sighed as he said, "James I am very sorry for your loss. I know your feelings, and nothing I can say can improve it. He's in a better place now, and my wife is in a better place.

Thank you, James, and thank you and your brother; if it weren't for you, I never would have seen my daughter again. I am forever grateful." He reached his arm out to shake my hand. I looked up and met his grip. I sadly responded, "Yeah, don't mention it, and yes, they are. Go to your daughter now. Best to you, Francis. Goodbye." He smiled at me one more time and went back. This would be the last time I spoke with Francis in person. Bates came over from the medical tent and hugged me. He rested his head on mine and said, "Walker, I'm so glad you made it out, man, my brother. I didn't know Thomas that long, but from what I saw, that's the best brother a man could ask for. I'm truly sorry, man; he saved our lives though, and that's an amazing act of courage if you ask me. Oh man, what a shitshow. I went to hell and back looking for you, you know." I chuckled slightly as I responded, "I know. I'm grateful to have a friend like you who would do anything for me." Bates rested his hand on my shoulder and shook it as he said, Well, come on, I couldn't let my redneck friend die; who's going to be left to lasso up the bad guys and fuck his sister?" Although I was sad, I couldn't help but laugh and say, "Fuck you, asshole, thank you, man; truly, thank you for coming back for me." Bates grinned and nodded. "Alright, I'll talk to you later.

Tent's calling me back, legs fucked, and shit hurts like a motherfucker. I've got to get the drugs, so I'm going to go. Get home safe, Walker. We'll be in touch." I shook his hand, and he was gone.

Captain Jacoway and McDonald approached me next. "I see they patched you up pretty well," said Captain McDonald as he let out a small chuckle. Captain Jacoway comforted me and said, "Walker, you survived what I can only imagine to be the hardest thing of your life. I'm very sorry. I'm sorry for your brother; he seemed like a good man, a good soldier; he died protecting the civilians, you, and his country—an honorable man at the least. I'm proud of both of you. You stopped an evil man from harming anyone else, and I couldn't be more proud of you for that. Another unit should be here soon; it's over, Walker; you're going home." And I cried a little bit more as I asked the captain, "Sir, do you think it was all worth it? I mean, the thousands of deaths of young men and innocent men. I mean, to die, and for what? What did we fight for? Thomas didn't die for country; he died saving my ass, just like when we were kids and my dad got too drunk and hunted me to beat my ass for the hell of it. Now that he's gone, what family do I have left?" Captain Jacoway looked at me and said, "James, you didn't fight for country; you fought to go home, be with your lady again, go home, kiss her, start a family, and make something of yourself. Your brother didn't die for nothing; he died giving you the chance to live your best life. James, it's not about what you're fighting for but who you are fighting for." Captain Jacoway and Captain McDonald shook my hand, let me be, and let me grieve privately. Finally, I could rest, and although the death of my brother will scar me forever, I felt relieved that I was done with this country and could finally see Leah again. The unit arrived about an hour later, and we loaded into the trucks. Thomas's covered body was placed in front of me. We loaded onto a plane, and I was on my way home before I knew it.

CHAPTER 16

GONE BUT NOT FORGOTTEN

My service to the war was over; I did my job, and now that I'm back in Georgetown, Texas, I feel like I'm on another planet. The sense of home is absent. I didn't stop until I got home, and once I swung open the door, I didn't see Leah anywhere. I checked upstairs and in the basement. I searched the garage and found nothing. I stopped when I entered the living room and wondered where she could have been. Suddenly I could hear singing coming from the backyard. I approached the screen door and saw Leah on her hands and knees, attending to her local garden that she had been working on since I left. She sang a song by Harry James: "Kiss me once, kiss me twice, and kiss me once again." I opened the screen door slowly, interrupted her, and finished off the lyrics, "It's been a long, long time." She gasped, stared forward, dropping the trowel out of her hands, and began crying. She stood up and turned around slowly, and once she saw me, the tears grew stronger.

Approaching her slowly I continued to sing, "Haven't felt like this before since can't remember when." She began laughing hysterically as tears continued down her face and she finished the lyrics after, "It's been a long, long time." She sprinted toward me and wrapped me in a big hug, kissed me with force and out of anger and fear slapped me across my face, " I thought you were dead, I didn't hear from you, I thought they were going to come by and tell me what happened to you." I grabbed her by the cheek and smiled as I said, "Well I had to come home, I made a promise to a very special lady, I'm back Leah, and I'm not going anywhere." She laughed and smiled again and pulled me in for another kiss. She asked me, "Wait, where is Thomas? "I turned my head down and hesitated to answer her question. After building the strength and holding back tears, I gave her the news: "He uh, he didn't make it; he sacrificed himself to save me; oh, Leah, I've been through hell, but holding you now, none of that matters; he died so I could come back to you, and now that I stand here, I have to ask." Dropping to one knee, I stared at her weeping eyes and asked, "Leah May, will you marry me? "She was shocked and tried wiping the tears from her eyes as she responded, "Of course I will, yes, yes." I stood up and kissed her again. The next couple of days were filled with passionate nights and spending time with her as I told her about the war and why I couldn't send letters. We spoke about D-Day, my capture, and Francis and Bates. She could not believe the horrors I suffered and was glad that I was finally back where I belonged.

Days passed, and I requested that Thomas's body be brought back to the house where we could bury him on the hill beneath the willow tree. Thomas's funeral would be held on a Saturday, and Leah was dressed in black. I dug a grave for Thomas and placed a cross at his head. I hung his tags and garrison cap on the cross as souvenirs of my fallen brother. I couldn't help myself and cried once more, as I still felt guilty for allowing my brother to die. I of course knew that there was nothing that I could have done, but the emotions you feel after watching your brother's life be taken are a feeling that you never overcome. Leah wrapped her arms around me, hugged me from behind, and reassured me, "Your brother was a good man, James; he's resting now, the pain is gone, and he watches over you from heaven. Your mother and father would be so proud of him and the sons you turned out to be.

They are all watching over you now." Composing myself, I nodded and said, "Yeah, I know. I'm going to miss you, Tom. Hey, remember when we were kids and your dumbass ran away? "I laughed, fighting the tears, remembering the nostalgia we shared as kids. "Pa made me look for you, and I lied, saying it was just us playing hide and seek. He was so angry when he found out you left with the neighbor's boy in town." I dropped to my knees in front of his grave again. "And those soldiers were walking through, son of a bitch Thomas; why couldn't it have been construction workers walking through? I reckon that would have been a better option for the both of us." I said my final goodbyes to Thomas and hugged Leah tight, knowing she was truly the only family I had left.

Months had passed, and on a Sunday, the time came to see where Leah and I would get married. It was a small service where Leah and I were at our local church. I had sent an invitation to Bates to see if he would show up. Bates walked through the door in his dress uniform and sat in the front row. With him, to my surprise, were Captain Jacoway as well as Captain McDonald. The pastor asked, "James, would you like to give your vows now? "I held onto Leah's hands tightly and told her, "Leah, growing up, I had nobody to trust except my brother. We grew into this strong bond for years, and I couldn't rely on anybody but him. My father was a drunk, and my mother passed away before we could make any sense of our lives. I was lost when I joined the Army, and I thought if I lost Thomas, I would have nobody left. At least that's what I thought until I met you; you brought the happiness back in my life, and I don't want to be with anyone but you. I love you, Leah, with all my heart, and I promise to love you forever and always." Leah gave her vows and we both shared a couple of seconds to cry and enjoy our lives at the moment, and the pastor permitted me to kiss the bride.

After the ceremony, I caught up with Bates and the captains. Captain Jacoway grabbed some medals from his pocket and handed them to me. The first one was a Purple Heart for being wounded at war and surviving Auschwitz. The second medal was another Purple Heart for

Thomas. Captain Jacoway said, "For his bravery and devotion to his country, James, this is for Thomas giving his life to save hundreds; I am very proud of both of you." I accepted the medals but knew this could never replace my brother's life. Captain McDonald came closer and said, "There is one more thing, Walker: you and your brother have earned the Medal of Honor; the President wants to meet you for a private ceremony; and you leave Tuesday." I shook both of their hands and said thank you and that I was incredibly happy to see them again. They returned to their vehicles, and I wouldn't see them until I got to Washington. Bates stayed afterwards to finally meet Leah, and they had a couple of drinks at the house. We indulged in conversation for hours, speaking of our experiences and how we were both nearly traumatized and lucky to be alive. "How's the leg holding up? "I asked as he rested his crutches against the table. "Healing I suppose, they gave me a purple heart too, but it doesn't mean nothing to me, not after what your brother did. They bring him back home? " I cracked open a beer and took a few sips as I looked out the window and pointed in his direction. "Yeah, we buried him out by the willow over there; I reckon some shade will do his hot head some good. You know, Bates, I can't help but think that if I got off Zeigler sooner or just let the bombs crush him in the debris alone, would Thomas still be here? " Bates lowered the bottle from his lips and sighed, knowing that there was nothing I could have done. "Come on, man, you know that's not true. If it weren't for him, Francis probably wouldn't have reunited with his daughter, and lives could have been lost. It was fate, and a good one at that. Your brother is a hero; you should be proud. Look, man, I have to go, but I'll see you in Washington, D.C., alright? " He polished off his beer, and before he left, he rested his hand on my left shoulder. "It was really good to see you, brother; I'll see you in a few days." Bates said goodbye to Leah and left.

Tuesday, August 7, 1945 came, and before we left for Washington, I started my day with coffee and read the newspaper. The headline read, "Nuclear Strikes against Japan." I called Leah over to come read what had happened, and she covered her mouth with shock. I had heard a man by the name of Robert Oppenheimer was put in charge of the Manhatten project that required the creation of nuclear weapons but we never thought it would actually be put to use. Leah and I were devastated that it came to such use of these weapons to end all of this. Finally, the war was over, and Japan had surrendered. Fortunately, no longer would innocent boys and men have to devote their lives to war; America had won. America will have suffered nearly four hundred thousand casualties by this time, and the reason to fight was misunderstood. Leah said, "Alright, come on, you have to pack; we will leave soon." I packed enough for a day's trip and left for Washington. We met with Bates and the captains when we arrived at the airport. We were driven to the White House and escorted to the East Room, where the ceremony would occur. As we sat in the chairs, we heard a group enter the room, and President Harry S. Truman stood in the center. We stood at attention, and Captain Jacoway rendered a salute. President Truman told us to take a seat and wanted to speak with me in private in the Oval Office. I followed him by his side, where he asked me to sit in the chair before the desk and had the secret service member shut the door. "So, I understand you're from Georgetown, Texas, is that right?" "Stiff as a log, I sat up straight and responded, "Yes, Mr. President, that is correct."

He sat in his chair and lit a cigar. "Well, it's nice to speak with someone else from the southern land; I'm from Missouri myself; listen, I heard what happened to you and my condolences to your brother; he seemed a good man; that's the other reason I wanted to speak with you; you're not the only one receiving a medal today." I sat there confused on what he was talking about, and he stood up and placed the cigar in an ashtray and went to the front of the desk. "Your brother has earned one himself; to give his life in the service of his country and to save many prisoners and you and your men from certain death is one of the highest acts of courage I have ever heard of. I want you to accept his award on his behalf." I was honored to have the opportunity to shine light on Thomas one more time and happily accepted the President's request. He spoke a bit more about what we went through and proceeded to the east room for the ceremony. President Truman said in his speech, "Lest we forget the brave men who dedicated their lives to an impossible cause. Sergeant James Walker did exactly that. He has earned the Medal of Honor for showing courage and bravery during the battles that took place in Germany and for saving countless lives alongside his men and his brother. Sergeant Walker neutralized a great threat to the civilian population and rescued, with the help of his men, the prisoners of Auschwitz. Our country is in great debt to you, James.

Thank you and your brother for your service and congratulations." President Truman walked behind me and wrapped the medal around my neck. After, he stood before me and delivered the medal that was for Thomas, he shook my hand. Privately, he whispered, "I know this medal doesn't mean much to you and it won't bring Thomas back, but if there is anything you need, just ask, ok?" Letting go of his hand, I stood there looking at the crowd and saw Leah smiling and clapping. I looked back at the president and smiled, knowing the one thing I would want now. Looking into his eyes, I said, "Mr. President, the only thing I want from now on is a normal life, to be with my wife and start a family." He gave a soft laugh and nodded. "You got it, son; we will do that for you; your parents and your brother are very proud of you, as am I. Safe travels home, James." The ceremony had ended, and we said goodbye to the President and I said my goodbyes to the Captains and Bates. I wouldn't see the captains after that day, but I always wrote letters when I could. As for Bates, we often kept in touch, we would make time to find each other again and hang out every other few months. Leah and I would have our first son about a year later. When deciding what to name him, we agreed rather quickly; his name was Thomas

Bates Walker. I stayed true to my desire to have a normal life and became a farmer, just like my pa before me.

We stayed in the farmhouse where we raised Thomas. I taught him everything I knew, just like my father had taught us: how to act, speak, and be respectful. I taught him how to use the plow and grow crops when he was old enough. Unsurprisingly, he was a smart lad with a good head on his shoulders. Leah stayed home and helped out when she could, and she liked to crochet when she could. She cooked meals and used a lot of ingredients from her garden. She was proud of the crops she grew. Nine months later, we had a little girl, and we named her after my mother. Her name was Betty Leah Walker; she was a beautiful and sweet child, just like her mother. My goal was to dedicate my life to protecting my family and providing for them. After witnessing what had happened to Francis's daughter, I would take the life of anyone who would seek to bring harm to my family. I was no longer a soldier fighting for my country; I was a proud father of two, a happy husband, and a successful crop farmer. I was home and Texas country again. This is where I belonged; nothing would take that away from me. It was a nice day, and after working in the fields, we decided to have a family picnic for lunch next to Thomas' grave. We packed some sandwiches and potato salad, and I brought the medals that Thomas had earned. Leah placed a blanket on the ground and set up the food. After eating, Leah took the children back inside to wash up. I stayed behind and talked to my brother for a little bit. "Farms going well, they suspect a good crop this year; Thomas is growing so big now; you should see him; he's damn near half my height, I reckon; Betty is getting into dolls and playing house; she's so cute when she acts like she's cleaning; she has a plastic broom that isn't picking up anything, but she enjoys playing; um, hey Tom, I just want you to know I miss you and think about you every day. I met the President awhile back and won the Medal of Honor, I don't know, which is not that exciting to me, but you earned it as well, and a Purple Heart. I'll leave it on your cross for you. I hope you, mama, and pa are doing okay, and it must be so beautiful there. I'll never forget what you did that day, and the lives you saved aren't going to be forgotten either.

I love you, Thomas, and I'll talk to you later. Bye for now." I rested my hand on his cross, and that's when I heard Leah yell from the back door, "Alright honey, come on back inside now, wash up." I turned to her and stared at her for a moment as how she said it brought back the nostalgia of when my mother would call Pa and I back for dinner. I smiled, looked up at the sky, inhaled deeply, and then yelled back, "Coming, dear, I'm coming." I turned to Thomas one more time and headed back inside. When I approached the door, I kissed Leah, held my children, and kissed their foreheads. This was the life I always wanted. Loving family was a life I was proud of, and I looked at them all and said, "I love you all very much; come on, let's wash up now." Long days in the field with my son caused us to become fatigued, but a hard- earned day was rewarding, and we knew that crop season was just around the corner.

END

Made in the USA
Monee, IL
11 November 2026

bdf6d638-e449-4152-bcb0-d6c6c2b45254R01